Space

Space

our final frontier

John Gribbin

This book is published to accompany the television series *Space* which was produced by the BBC and was first broadcast in 2001.

Creative Director: John Lynch
Executive producer: Emma Swain
Series producer: Richard Burke-Ward

Published by BBC Worldwide Limited,
Woodlands, 80 Wood Lane, London W12 0TT

First published 2001
Copyright © John Gribbin 2001
The moral rights of the author have been asserted

All rights reserved. No part of this book may be reproduced in any form or by any means, without the permission in writing from the publisher, except by a reviewer who may quote brief passages in a review.

ISBN: 0 563 53713 2

Commissioning Editor: Joanne Osborn
Project Editor: Helena Caldon
Academic Consultant: Dr Margaret Penston
Design concept: The Attik
Art Direction: Pene Parker, Lisa Pettibone
Art Editor: Kathy Gammon
Design: Bobby Birchall, DW Design, London
Picture Research: Carmen Jones and Miriam Hyman
Illustrations: Mark McLellan and Kevin Jones Associates

Set in Albertina MT and Gill Sans
Printed and bound in Great Britain by Butler & Tanner Ltd, Frome
Colour separations by Radstock Reproductions Ltd, Midsomer Norton
Jacket printed by Lawrence-Allen Ltd, Weston-super-Mare

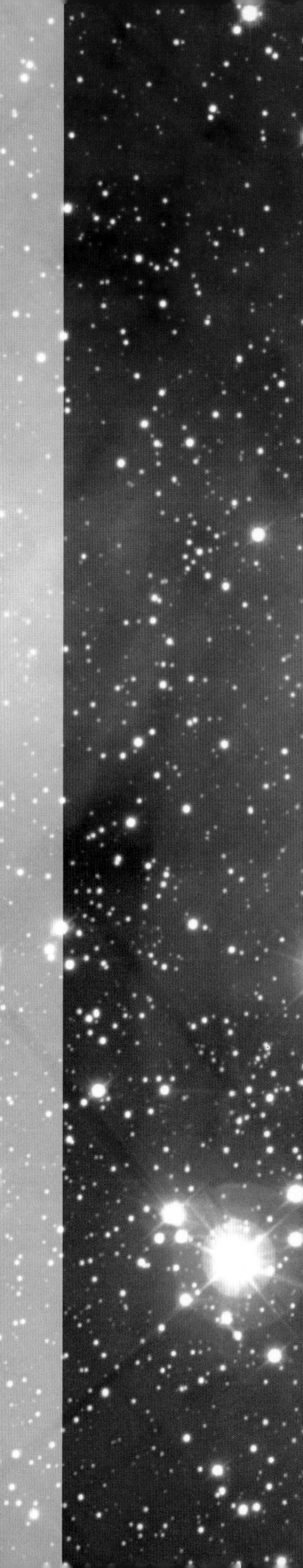

Contents

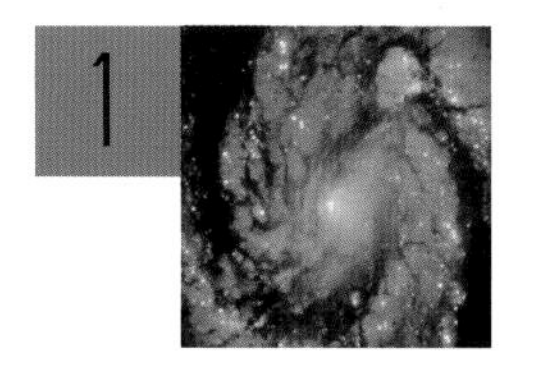

1

ACROSS THE UNIVERSE

Astronomers only began to understand the way the Universe works once they developed techniques to measure distances across the Universe. Without a knowledge of distances, they had no way to measure the sites and brightening of stars and galaxies.

2

THE FATE OF THE UNIVERSE

Will the Universe expand forever? Or will it one day collapse into a black hole? Nobody knows the answer, but 21st century cosmologists at least know where to look for it.

3 MAKING CONTACT

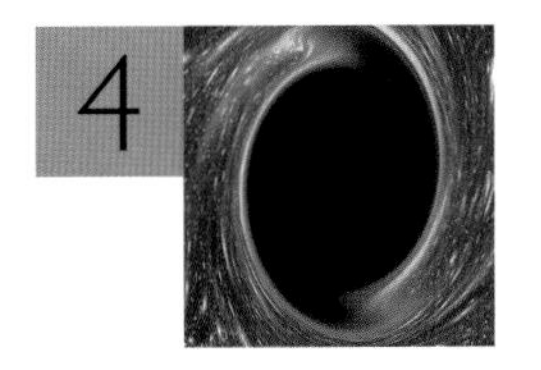

4 OTHER WORLDS

1

ACROSS THE UNIVERSE

1.1 STEPPING STONES TO THE UNIVERSE

The travellers in *Star Trek* boldly go where no one has gone before to explore the final frontier – space. No human has yet visited any object outside our own Solar System, but that has not stopped us exploring new worlds – at long range, using telescopes on the surface of the Earth and satellite-based instruments orbiting above the atmosphere. The data from these observations are then compared with what we can infer about stars and galaxies from the laws of physics. In astronomy, theory and observation always go hand in hand – a theory about stars is useless without observations to test the predictions of the theory, and observations of a startling new phenomenon remain a mystery until they can be understood within the frameworks of a theory about the Universe. Together, theory and observation can take us on a journey to the furthest reaches of the Universe, and back in time to when the Universe was born.

Previous page. A spiral galaxy, like the Milky Way in which we live.

MAKING MAPS OF SPACE

Astronomers are interested in the evolution of stars and galaxies (how these objects are born, live and die), and in tracing the origin and ultimate fate of the entire Universe. Putting this type of knowledge together with a knowledge of the distances between cosmic objects enables astronomers to achieve an understanding of their domain similar to a naturalist's understanding of the world – the equivalent of combining biology with geography, and finding out what different kinds of creatures live in different parts of the world.

By studying the light emitted by stars and galaxies, astronomers are able to find out what different kinds of objects exist in different parts of the Universe. But they also need to measure the distances to cosmic objects, so they know where they are in relation to one another. And how can the distances to stars and galaxies that we have no hope of ever visiting, even with an unmanned space probe, be measured? It sounds like an impossible task, but astronomers have found 'stepping stones' that effectively take them from Earth to the furthest reaches of the Universe.

It's all Done with Triangles

As the Chinese proverb says, the longest journey begins with a single step; and the geographical exploration of the Universe starts with a simple piece of geometry involving triangles.

The first step into the Universe uses exactly the same kind of surveying techniques used here on Earth to measure the distances to distant objects (such as mountains) without actually having to go there. The idea itself is not new, but with the aid of new instruments here on Earth and satellites orbiting above the Earth it reaches further than ever before.

It all depends on the geometry of triangles. If you know the length of one side of a triangle (the base) and you can measure the angle each of the other two sides makes with the base, then it is a simple matter to calculate how far it is from the base of the triangle to

1. Human beings have only visited our nearest neighbour in space, the Moon.

2. Exploration of deep space relies on remote observing using instruments such as radio telescopes.

the opposite tip. The process is called triangulation, for obvious reasons.

The trouble with triangulation is that you need a longer baseline to measure the distances to more distant objects.

The Importance of Parallax

Triangulation is not restricted to measuring distances on Earth – it works very well for measuring the distance to our nearest neighbour in space, the Moon. If one observer sees the Moon directly overhead, for example, while another observer, standing on what is the horizon for the first observer, also measures the angle it is in the sky, then it is easy to work out the distance to the Moon – about 384,000 km – from the geometry of triangles.

This is possible because the Moon appears in a different part of the sky to the two observers. It is exactly the same as the way you can make your index finger jump across a distant background if you hold your arm straight out in front of you, point your finger upwards, and look at it with each eye closed in turn. The slightly different views from your two eyes give you different perspectives on your finger; and the slightly different views from two observatories give different perspectives on the Moon. The effect is called 'parallax', and for our two observers it shifts the apparent position of the Moon nearly twice the diameter of the full Moon on the night sky.

But the distant stars are so far away that the background pattern looks the same from anywhere on Earth, and this makes it convenient to measure parallax by measuring how far the Moon (in this case) seems to shift against the fixed background of stars.

Beyond the Moon

Triangulation and parallax have also been used to measure the distances to the nearest planets, Venus and Mars. This is much harder than measuring the distance to the Moon, because the planets are much further away. It involves making observations from opposite sides of the Earth at the same time, then calculating the geometry of a very tall, thin triangle.

The parallax of Mars was determined accurately in 1671, when the French astronomer Jean Richer led an expedition to French Guyana (in South America) to measure the position of Mars against the background stars at a certain time on an appointed night (actually several nights, to allow for cloud).

1

MAPPING THE EARTH

Triangulation is a technique that can be used to measure distances to distant objects. A map-maker can plot the location of a hill on a map by first measuring out a baseline, perhaps a kilometre or so long then, using small telescopes (called theodolites), measuring the angle between the baseline and the hilltop – from both ends of the baseline. Using these angles, and knowing the length of the baseline, the map-maker can now calculate the lengths of the other two sides of the triangle, and thus how far the hill is from the baseline. The distance from one end of the first baseline to the hilltop can be used as a new baseline to measure other distances. You can use this process repeatedly to work your way across the surface of the Earth. Indeed, although the technique has now been superseded by satellite mapping, triangulation was used by nineteenth-century surveyors to map India, starting at the southern tip and working their way northward to the Himalayas.

On the same nights and at the same times, back in Paris, the Italian-born astronomer Giovanni Cassini also made observations of the position of Mars against the background stars. When Richer's expedition returned, the two teams compared notes and calculated the distance to Mars.

Law-abiding Planets

These measurements were particularly important, because they made it possible to work out the geography of the entire Solar System.

The laws which describe the motion of the planets around the Sun were described early in the seventeenth century by Johannes Kepler, and explained by Isaac Newton with his theory of gravity. They state that, if planet A is twice as far from the Sun as planet B, then the orbital period of planet A (the time it takes to go round the sun: its 'year') is a certain multiple of the orbital period of planet B.

Astronomers thus had to measure at least one planetary distance directly in order to put real numbers into the equations, even though they already knew the orbital periods for the planets. By measuring the distance to just

The observations that led to the first measurement of the distance to Mars were made in French Guyana, near the site used by the European Space Agency to launch its Ariane rockets.

2

1. Triangulation and the parallax effect.

2. Once astronomers had determined the sizes of galaxies, they could even use triangulation to estimate the distances between them from how small the galaxies look on the sky.

1

3

1 and 3. Johannes Kepler discovered the laws of planetary motion by studying the orbit of Mars (right).

2. Kepler's key discovery was that planets in their orbits trace out equal areas in equal times.

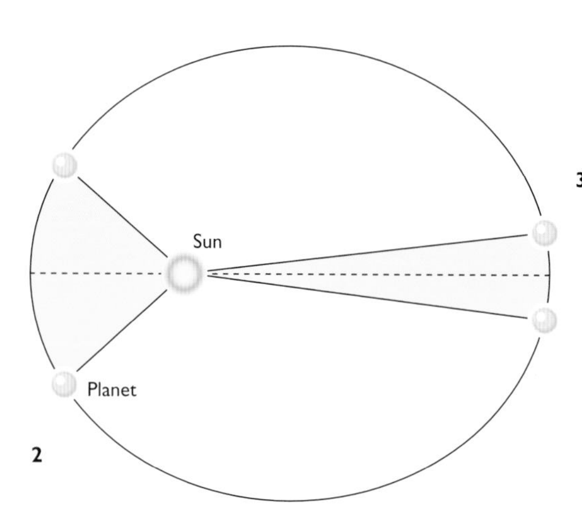

2

Venus and Mars, they were able to calculate the distance from the Sun to each of these planets. Once they knew these distances, they were able to use Kepler's laws to calculate the distance from the Sun to all the other planets in the Solar System – including the Earth. In addition, they could use Newton's laws to calculate what the mass of the Sun must be to hold the planets in these orbits through gravity.

By the end of the seventeenth century, astronomers were able to calculate the distance from the Earth to the Sun fairly accurately. The observations have been improved since then (we can even measure the distance to Venus directly, by bouncing radar signals off it), and the distance from the Earth to the Sun is now known to be 149.6 million kilometres (almost 4000 times the distance around the equator of the Earth). But even 200 years ago, the calculated distance was 140 million km – an error of less than 7 per cent compared with the modern figure.

STEPPING STONES TO THE STARS

It takes the Earth 12 months to orbit the Sun once. The radius of the Earth's orbit – the distance from the Earth to the Sun – is roughly 150 million km. This distance is called the Astronomical Unit, or AU. It is vitally important in astronomy, because it provides a new baseline with which to measure parallaxes for more distant objects – the nearest stars.

At intervals 6 months apart, the Earth is at opposite sides of a diameter measuring 2 AU (about 300 million km). This is such a long baseline that in photographs of the night sky taken 6 months apart a few of the stars seem to have shifted their position slightly, because of the parallax effect. But the shift is very slight, because the stars are so very far away. To give you some idea how small the effect is, in the 1830s the first star studied in this way (known as 61 Cygni) was found to have a parallax shift of just 0.31 seconds of arc. (There are 360 degrees in a circle, 60 minutes in a degree, and 60 seconds in a minute.) In comparison, the full Moon covers 30 seconds of arc on the sky. So the apparent shift in 61

Mount Everest is named after Sir George Everest, the surveyor who led the team that mapped India in the nineteenth century.

4

4. We live in a spiral galaxy like this one (M65) containing hundreds of billions of stars like the Sun.

1

Cygni as the Earth goes round the Sun is equivalent to about one-six-thousandth of the apparent diameter of the Moon.

The distances to the stars are so great that astronomers had to invent new units with which to describe them. If you were so far away from Earth that the distance between the Earth and the Sun (the radius of the Earth's orbit, 1 AU) covered just one second of arc on the sky, then you would be one parsec away from Earth (the term 'parsec' is a contraction of 'parallax second of arc'). A parsec is just over 30 million million kilometres, a number so big that it is hard to visualize – but you can look at it in terms of the speed of light. Light travels at just under 300,000 km per second, and so covers 9.46 million million km in a year, a distance known as a light year. So a parsec is 3.26 light years. Converting the parallax measurement into distance, we find that 61 Cygni is 3.4 parsecs away, or just over 11 light years from us. And, amazingly, this makes it one of the closest stars to our Sun.

Stars Like Dust

When you look up at the sky on a dark and cloud-free night, it seems to contain countless numbers of stars, and poets have waxed lyrical about the view. But the human eye is not very sensitive to faint light: even under perfect conditions, with no Moon or cloud, and far from city lights, the most you can see at any one time is about 3000 stars. Under more ordinary viewing conditions, you are lucky to see a thousand.

The true numbers of stars in the sky only began to be appreciated at the beginning of the seventeenth century, when Galileo Galilei turned his telescope onto the night sky. He found that, what seemed to be a faintly glowing cloud of light was actually a myriad of individual stars, each too faint to be seen by the unaided human eye. He announced his discoveries in a book, *The Starry Messenger,* which was published in 1610.

At that time there was no accurate way to estimate the distances to the vast majority of these stars. Until very recently, only a few stellar distances had been measured directly by parallax. By the end of the nineteenth century, just 60 stellar distances had been measured in this way. At the end of the twentieth century, the situation improved dramatically when the HIPPARCOS satellite, orbiting clear of the obscuring influence of the Earth's atmosphere, measured the distances to a large number of stars with unprecedented accuracy. It pinned down the parallaxes of more than 100,000 stars, to an accuracy of 0.002 seconds of arc. But even this impressive achievement gives the distances to less than one-millionth of the total number of stars in the Milky Way, taking the range of directly measured stellar distances out to a few hundred parsecs (about a thousand light years).

Colour, Brightness and Distance

So even with the aid of satellites like HIPPARCOS, astronomers still need other techniques to measure distances to stars outside our local region of space. The most important of these techniques is called the 'moving cluster' method. It gives the distance to a large group of stars, called the Hyades Cluster. These stars are about 40 parsecs (130 light years) away from us, and all move as a group through space (HIPPARCOS has, now, also confirmed the distance to this cluster). Because this cluster contains hundreds of stars with different colours and brightnesses, the fact that they are all at the same distance away helps astronomers to understand how brightness is related to the colour of the light emitted, in subtle ways. The colour of a star has no relation to its distance: it's the brightness that tells us how far away it is. Then, when they

2

1. The satellite HIPPARCOS, shown here being tested prior to launch, measured the distances to the stars with unprecedented accuracy.

2. The Hyades star cluster and the smaller cluster known as the Pleiades.

THE STARS ARE SUNS

The Sun is a star or, to put it another way, the stars are suns. Scientists only began to appreciate this in the seventeenth century, and even then they found it hard to believe. Isaac Newton, for example, calculated that if the star Sirius is really as bright as our Sun, then to appear as faint as it does in the sky it must be a million times farther away from us than the Sun. Newton was more or less right, but he was so discomfited by this calculation that he never published it during his lifetime (it appeared in a book, *System of the World*, published in 1728, the year after Newton died). Yet Sirius is actually one of our nearest stellar neighbours, only 2.67 parsecs (8.7 light years) away.

We now know that the Sun is a very ordinary star. It is neither particularly large nor particularly small, not unusually hot nor unusually cool, and it is roughly halfway through its life. This is disappointing for people who like to think that there is something special about our place in the Universe. But it raises a much more exciting possibility. If the Sun is an ordinary star, then it seems likely that other stars like the Sun will have families of planets orbiting them like the Solar System. Could it be that as well as being ordinary in every other way, the Sun is quite typical in having a family of planets that includes a home suitable for life? If so, there may be literally billions of other Earths out there in the Milky Way Galaxy.

SPECTROSCOPY: THE KEY TO ASTRONOMY

The single most important tool of astronomers is the ability to analyze starlight and discover what stars are made of. This depends on the fact that atoms of any particular chemical element radiate energy (if they are hot) or absorb energy (if they are cold) at very precise wavelengths in the rainbow spectrum of light. Each element produces its own distinctive set of lines in the spectrum when radiating or absorbing energy, yielding a pattern similar to a barcode. And, like a barcode, each pattern is unique.

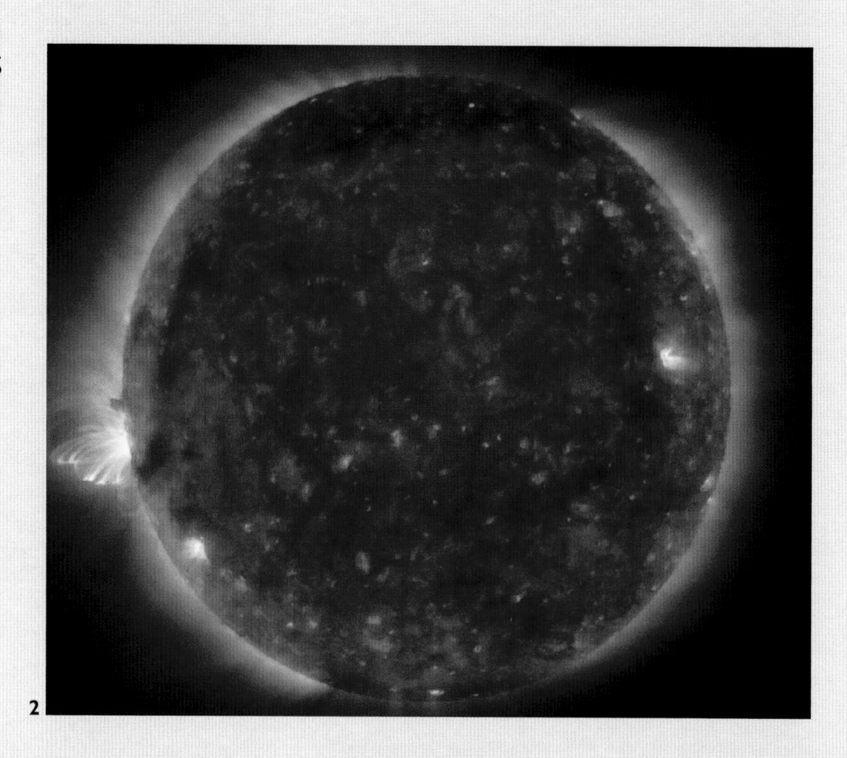
2

1

The Flame Test

We know which spectroscopic 'barcode' corresponds to a particular element because the light emitted by elements has been studied using simple flame tests. A sample of a known element (perhaps a piece of copper wire) is heated (often using a simple Bunsen burner) and the light it radiates when it is heated is passed through a triangular prism. This spreads the light out and produces a pattern of lines that is unique to that element. By repeating this test with many substances, a huge library of patterns has been built up, and an unknown substance can be identified by examining its spectroscopic pattern and comparing it with the patterns in the library.

Spectroscopy was invented in the middle of the nineteenth century, but there were many gaps in this library of knowledge, which still needed much investigation to fill. The first person to notice that light from the Sun, when passed through a prism to make a spectrum, contained many distinct lines was the British physicist William Wollaston, in 1802. But he had no idea what they were. In 1814, the German Josef von Fraunhofer counted 574 lines in the spectrum produced by light from the Sun, and discovered many of the same lines in light from the stars. But the person who explained that these lines were caused by the presence of different elements in the atmospheres of stars was Gustav Kirchoff. He pioneered the basic principles of scientific spectroscopy in collaboration with Robert Bunsen in Germany at the end of the 1850s.

Secrets of Sunlight

Spectroscopic studies of the light from the Sun's atmosphere, obtained during an eclipse in 1868, showed a distinctive pattern of lines which did not correspond to any known element. The British astronomer Norman Lockyer concluded that there must be an element in the Sun that had never been found on Earth, and gave it the name helium, from helios, the Greek word for the Sun. Helium was actually identified on Earth in 1895, and Lockyer received a knighthood (partly as a result of his famous prediction) in 1897. Spectroscopy had actually found an element in our nearest star before it had been found on Earth.

Moving Stars

There is one other vitally important use of spectroscopy in astronomy. Although the lines corresponding to a particular element are always produced at the same distinctive wavelengths, if the object making the lines is moving, the whole barcode pattern is shifted across the spectrum. If the object is moving towards us, the lines are shifted to shorter wavelengths. Because blue light has shorter wavelengths than red light, this is called a 'blueshift'. Similarly, if the object is moving away from us, the pattern is shifted towards longer wavelengths, a 'redshift'. This is known as the Doppler effect, and it enables astronomers to measure how fast stars are moving through space, how fast galaxies are rotating, and how fast stars in binary systems (where two stars orbit around each other) are moving in their orbits. The last application is particularly useful, because it is a key (along with the law of gravity) to measuring the masses of the stars involved.

So spectroscopy tells us what stars are made of, how fast they are moving, and what mass they have. Without spectroscopy, there would be little more to astronomy than making pretty patterns called constellations out of the arrangement of stars on the night sky.

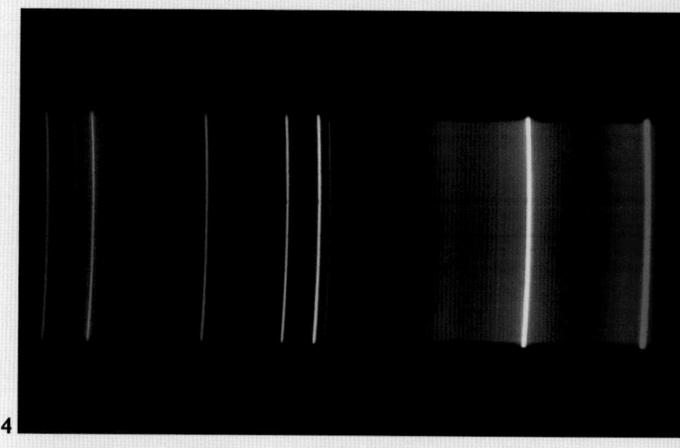

4

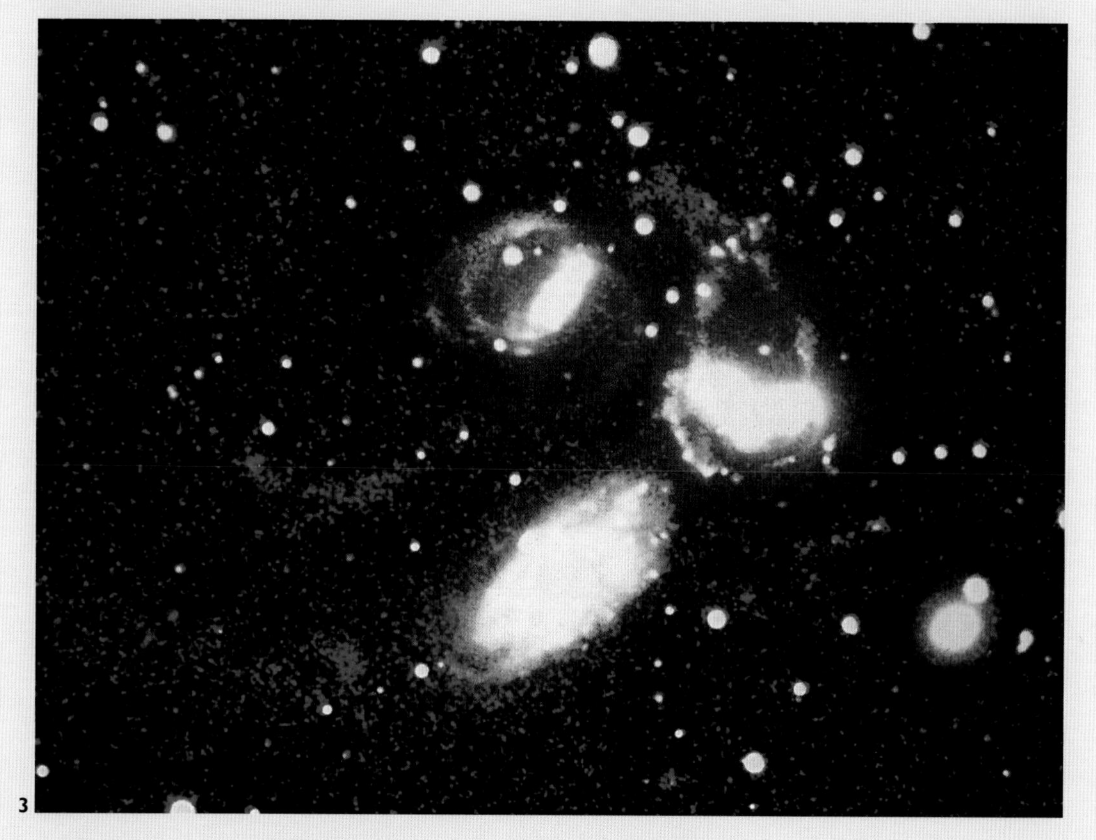

3

1. A bunsen burner being used to heat copper wire in a flame test

2. Solar corona - an ultraviolet image of the sun's outer atmosphere taken by the SOHO satellite. On the left is a prominence being absorbed after a solar flare event.

3. The Stephen's Quartet group of galaxies and NGC 7320. Colour coding shows the different red shift values of the quartet members.

4. The distinctive fingerprint of coloured lines is clearly shown here in the light from hot Helium gas.

TOPIC LINKS

1.1 Stepping Stones to the Universe
p. 17 Colour, Brightness and Distance

2.1 The Big Bang
p. 81 Galaxies on the Move

3.1 Life and the Universe
p. 140 The Big Surprise
p. 150 Seeking Signs of Life

1

2

see a star with the same colour as one of the types of Hyades stars, they can estimate its distance by comparing its brightness (or faintness) with the Hyades star. Crucially, the subtle differences in the colour of the stars involved are revealed by a technique called spectroscopy – probably the single most important tool used by astronomers.

The result of applying such methods is that we now have a clear idea of distances between stars, and also of their sizes. The distance from one star to even its nearest neighbours is usually tens of millions of times its own diameter (except, of course, for systems where two or more stars orbit around one another). For example, the Sun has a diameter of 1.39 million km (which is typical for a star during the main period of its life). If the Sun were the size of an aspirin, on this scale the nearest star would be another aspirin 140 km away. The distances between stars are absolutely enormous – even compared with the huge sizes of the stars themselves.

AN ISLAND IN SPACE

By using every possible technique for measuring distances to stars, astronomers have been able to map the collection of stars in which we live, an island in space called the Milky Way Galaxy, or just the Galaxy. This is rather like trying to map a forest from the inside by working out the distances and relative positions of the trees in all different directions. The process is aided by the fact that, in parts of the Galaxy, there are great clouds of gas and dust between the stars. These clouds contain large amounts of hydrogen, which can be detected by radio telescopes.

The overall shape of our Galaxy is a flattened disc containing hundreds of billions of stars, all more or less the same as our Sun, with a diameter of about 28 thousand parsecs (28 kiloparsecs). The disc is only 300 parsecs thick at its outer regions (roughly 1 per cent as thick as its width), but it has a bulge in the middle measuring 7 kiloparsecs across and 1 kiloparsec thick. If we could view our Galaxy from the outside it would look rather like a huge fried egg.

Surrounding the whole disc is a halo of about 150 known bright star systems called globular clusters. Each globular cluster is a ball of stars containing hundreds of thousands, or even millions, of individual

If you take a deep breath, you will have more molecules of air in your lungs than there are stars in all the galaxies in the visible universe put together.

3

1. An impression of the number of stars in our galaxy is given by this picture of part of the Milky Way.

2. The Milky Way has been mapped using star clusters.

3. Our Milky Way Galaxy is a flattened disc of stars embedded in a halo of globular clusters.

1. The Hooker telescope – used in working out the size of our galaxy.

stars, so close to one another that there may be 1000 stars in a single cubic parsec of space. From the way stars move, astronomers also infer that there is a great deal of dark matter (material that has mass but cannot be detected directly ▷ p. 96) surrounding the whole Galaxy and holding it in a gravitational grip.

Stars in Spiral Patterns

Viewed from above, our Galaxy has a distinctive structure, with bright trails of stars, called 'spiral arms', twining outwards from the central bulge. This is a very common feature of disc galaxies like the Milky Way, so much so that they are sometimes referred to as spiral galaxies. The most important distinction between the central bulge and the disc proper, however, is that the stars in the bulge (and the globular cluster stars in the halo surrounding the Galaxy) are all old stars. They are perhaps 12 billion years old and are known, for historical reasons, as Population II stars. There is also very little gas or dust in the bulge. The disc, where the spiral arms twine outwards, contains gas and dust and some old stars, but also middle-aged stars and very young ones, which are known as Population I stars (the Sun is a Population I star). New stars are still being formed in the disc all the time.

All the stars in the disc, together with the gas and dust, orbit around the centre of the Galaxy. But the disc does not rotate as if it were a solid object (the way a CD rotates as it is played). Each star moves independently – just as each planet in our Solar System orbits around the Sun independently – and the stars closer to the centre move faster than those near the edge of the disc. The Sun is travelling at a speed of about 250 km per second in its own orbit around the centre of the disc, carrying our Solar System with it; but the Galaxy is so large that even at this speed it takes our Solar System about 225 million years to complete just one orbit, a journey it has made about 20 times since it was born some 4.5 billion years ago.

The Sun and its family of planets orbits the Galaxy at a distance of about 9 kiloparsecs from the centre, two-thirds of the way out to the edge of the disc, on the inside edge of a feature known as the Orion Arm. We are not in the centre; there is nothing particularly special about our place in the Milky Way Galaxy.

A MATTER OF PERSPECTIVE

The size and shape of the Milky Way Galaxy were only really described properly in the 1920s. Before then, most people thought that the stars they could see in the sky made up the entire Universe – everything there was to see. But as well as revealing a myriad of stars in the Milky Way, telescopes also showed up faint patches of light in the sky, fuzzy blobs called 'nebulae'. At the same time that astronomers started to appreciate and understand the geography of the Milky Way, some of them began to wonder whether these nebulae might be other islands in space, galaxies like the Milky Way, but so far away from us that the light from the stars they contained only added up to a faint patch of light like a little cloud on the night sky. This suggestion caused a fierce debate among astronomers at the time, because it would mean that the other galaxies were at enormous distances from us, hundreds or even thousands of kiloparsecs away. This was hard to accept when astronomers had only just discovered that the Milky Way itself was several tens of kiloparsecs across, bigger than anything previously imagined. The other possibility was that the nebulae were indeed glowing clouds of gas within the Milky Way itself, between the stars.

The only way to find out whether the nebulae were actually galaxies was to identify individual stars within them and measure their distances directly. They would be too far away for triangulation to work; but by the 1920s astronomers knew that some kinds of exploding star (called novae) all have about the same brightness, while another kind of star (called 'Cepheids') have a brightness that can be inferred from their other properties. If you know the true brightness of a star, it is easy to work out how far away it is by measuring how bright it appears. So if the astronomers could identify novae and Cepheids in nebulae, they would be able to work out roughly how far away they were.

2

2. Supernova 1987A, a huge stellar explosion photographed in March 1987.

If each star were represented by a single grain of rice, a scale model of the Milky Way galaxy would just fit into the gap between the earth and the Moon.

It turned out to be just possible to make the crucial measurements for stars in some of these nebulae in the 1920s, using what was then the best telescope in the world. The telescope (still in use today) has a 100-inch diameter mirror, and is called the Hooker Telescope, after the benefactor who paid for it. It is located on top of Mount Wilson, near Pasadena, in California.

Beyond the Milky Way

The astronomer who made the crucial measurements was Edwin Hubble. He identified both Cepheids and novae in nebulae that are now known to be the closest galaxies to the Milky Way.

But it turned out that not all of the nebulae were other galaxies. Some of them really were clouds of gas and dust within the Milky Way, and these objects play an important part in the life cycles of stars and the origin of planetary systems like the Solar System.

In order to avoid confusion, astronomers kept the name nebulae for the clouds within the Milky Way, and used the term 'galaxy' to refer to the great star systems beyond the Milky Way.

Even with the 100-inch Hooker telescope, it was very difficult to make the observations needed to calculate the distances to galaxies. When Hubble first began to make

1. A galaxy like the Milky Way is made up of a central bulge of stars surrounded by a thinner disk.

2. (opposite) The Sun and the Moon both look the same size from the Earth even though the Sun is really much bigger than the Moon. In the same way, galaxies look tiny on the sky because they are so far away.

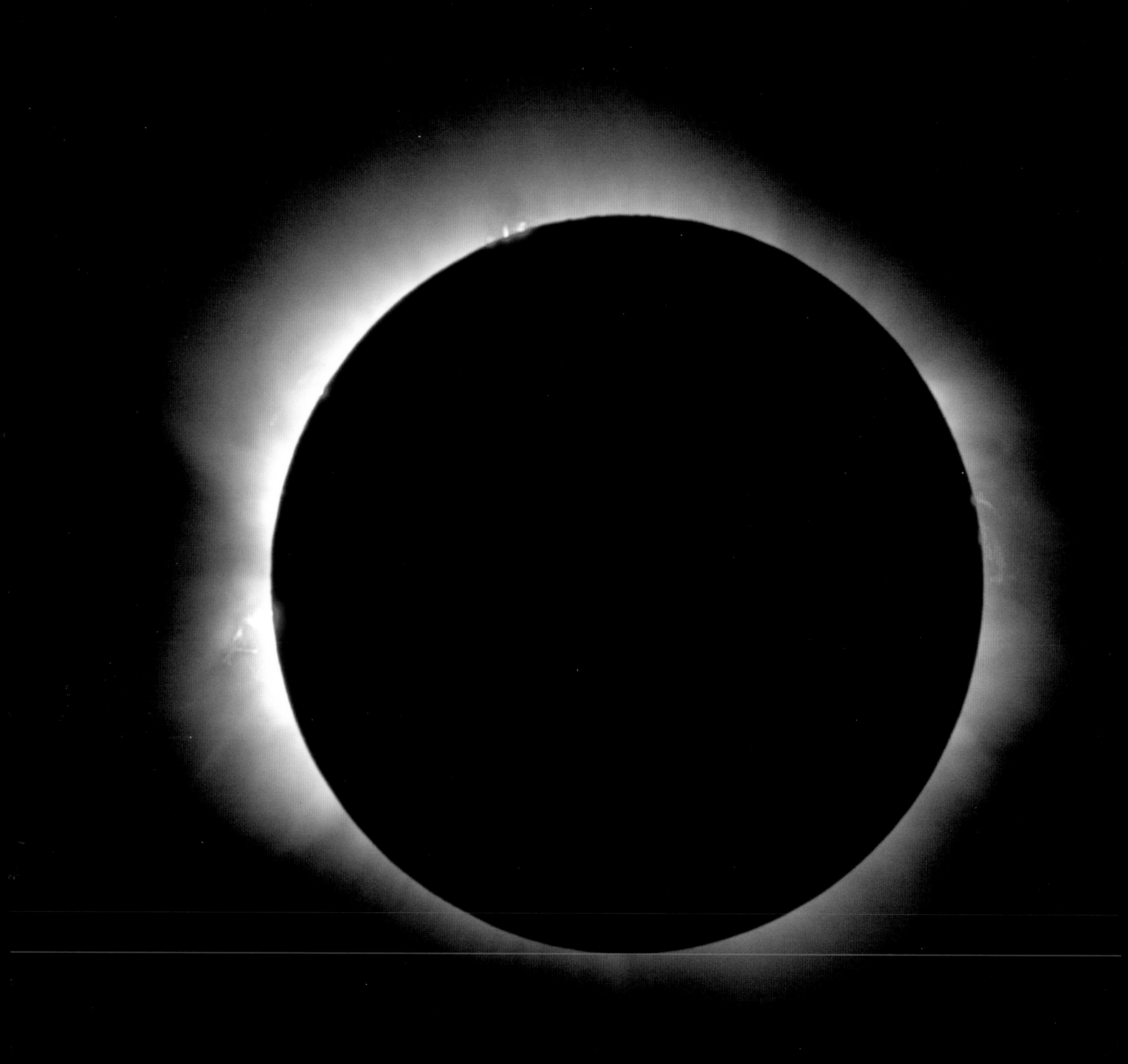

measurements of the distances involved, he found that, although the other galaxies did indeed lie beyond the Milky Way, they did not seem to be as big as ours. It's all a matter of perspective. One of the few things we can actually measure is the area a galaxy covers on the sky. A small galaxy close up will cover the same area as a big galaxy further away, in just the same way that if you hold a small model of a cow up in front of your eye, it looks as big as a real cow on the other side of a field. In the same way, the Moon completely covers the Sun during a total solar eclipse because, although the Sun is almost 400 times bigger than the Moon, it is also almost 400 times further away and so it appears to be the same size.

As telescopes got better and better, astronomers were able to measure the distances to other galaxies more and more accurately. They used many different stepping stones, not only Cepheids and novae, but also comparisons of the brightness of things such as globular clusters in one galaxy with those in another. After more than half a century of effort, they found that the galaxies were about 10 times further away than Hubble had thought, so it followed that they must be that much bigger than he had thought in order to look as large as they did on the sky.

However, Cepheids and novae are still the best indicators of distance. In the 1990s, using Cepheid distances obtained from the Hubble Space Telescope, a team at the University of Sussex finally showed that the Milky Way Galaxy is an average galaxy of its type (if anything it's a little bit smaller than the average disc galaxy in our part of the Universe). Like our position in it, there is nothing special about the Milky Way Galaxy.

Putting Galaxies in Perspective

The result of all these efforts is a clear understanding of the sizes of galaxies and the distances between them.

As well as disc (spiral) galaxies like the Milky Way, there are much larger, elliptical galaxies, which do not have a disc or spiral shape, but are ellipsoidal (like a rugby ball). These are thought to have been built up by a kind of cosmic cannibalism, from mergers between disc galaxies.

There are also smaller elliptical galaxies (resembling the globular clusters ▷ pp. 21-2) and small irregular galaxies which have no distinct shape. The largest elliptical galaxies contain several thousand billion stars. Disc galaxies, such as the Milky Way, have diameters of a few tens of kiloparsecs and contain a few hundred billion stars.

Galaxies are much closer together, relative to their own size, than the stars are to one another. Again, it's a matter of perspective. If we adapt the aspirin analogy to galaxies, and represent the Milky Way by a single aspirin, we find that the nearest large disc galaxy to us, the Andromeda Galaxy, would be represented by another aspirin just 13 centimetres away. And only 3 metres away we would find a huge collection of about 2000 aspirins, spread over the volume of a basketball, representing a group of galaxies known as the Virgo Cluster. On a scale where a single aspirin represents the Milky Way Galaxy, the entire observable Universe would be only a kilometre across, and would contain hundreds of billions of aspirin. In terms of galaxies, the Universe is a crowded place.

1. Most galaxies are grouped together in associations that are called clusters. This is the centre of the Virgo cluster of galaxies, which is a crucial stepping-stone into the Universe at large.

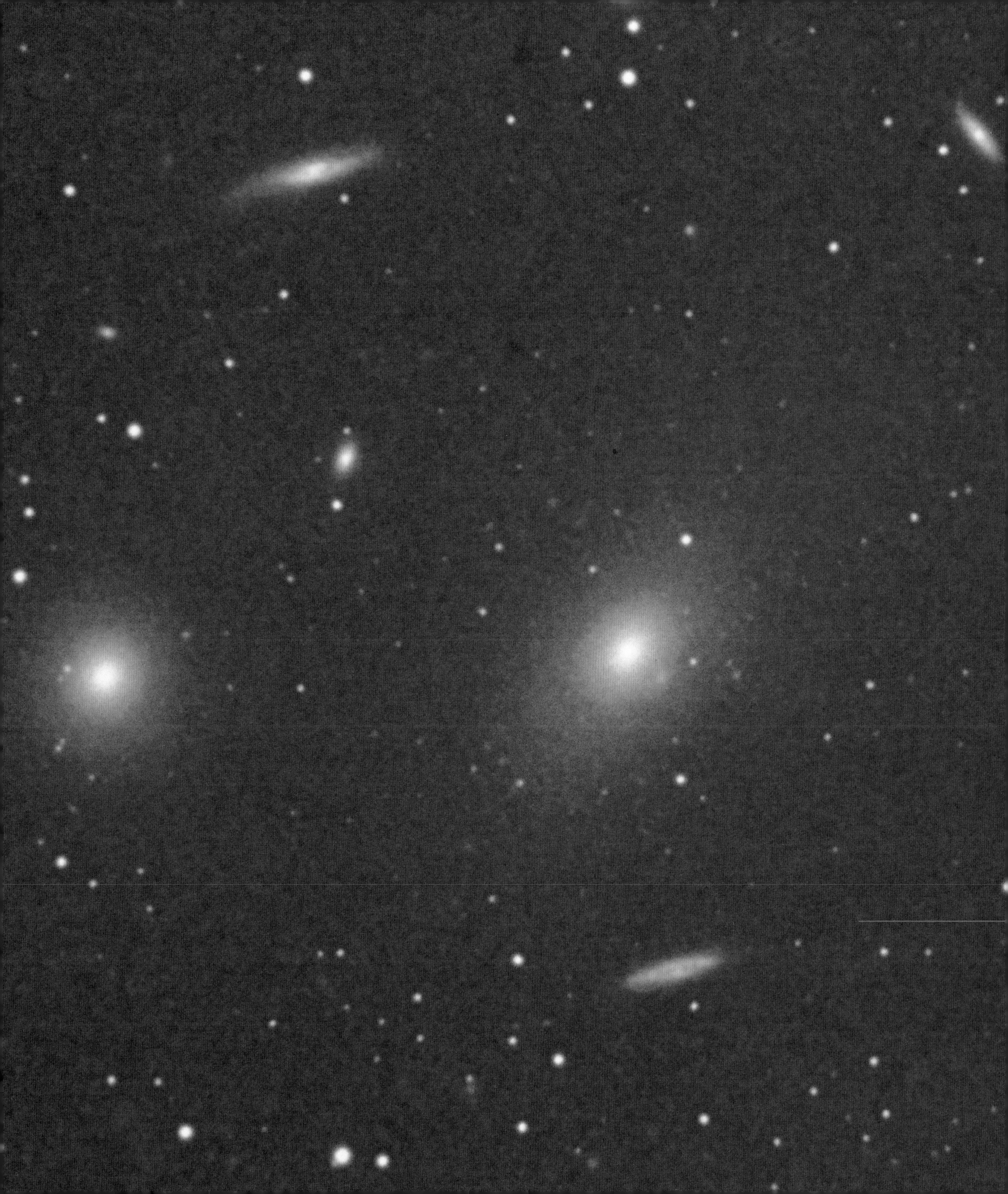

THE CEPHEID SCALE

Variable stars called Cepheids are the key to measuring distances across the Universe. Each Cepheid goes through a very regular periodic change in brightness; some have periods as short as a day, some take 50 days or so, others run through their cycle between these extremes. While they are interesting in themselves, their importance in astronomy derives from the fact that the Cepheids provide a vital stepping stone to the Universe – they give us accurate distances to the nearest galaxies.

2

Cepheids are large, yellow stars which have an intrinsic brightness between 300 and 26,000 times that of the Sun. They are typically between 14 and 200 times the size of the Sun. The change in brightness of a Cepheid is associated with a rhythmic pulsation, as if it is breathing in and out. A Cepheid is faintest when its outer layer has expanded to its maximum size and has cooled down, and brightest when the outer layer has contracted to its minimum size and is hotter.

First Steps

In the early years of the twentieth century, Henrietta Swan Leavitt was working at Harvard College Observatory, studying Cepheids in a star system called the Small Magellanic Cloud (which we now know to be a small irregular galaxy in orbit around the Milky Way). She noticed that the brighter a Cepheid is, the more slowly it goes through its cycle.

This showed up clearly for Cepheids in the Small Magellanic Cloud because the cloud is so far away from us that the distance from one side of the cloud to the other is such a small fraction of the distance to the cloud, that all the stars in it can be regarded as being at the same distance from us.

So if one Cepheid star looks twice as bright as another, it really *is* twice as bright, not just closer to us.

Stellar Street Lights

Leavitt's discovery of the relationship between brightness and period for Cepheids meant that they could be used to measure relative distances within the Milky Way. The period–luminosity relationship might tell you that one Cepheid is intrinsically twice as bright (say) as another, and comparing their apparent brightnesses on the sky would then reveal their relative distances from us. But in order to use this information to measure actual distances across the Milky Way, the distances to at least a few Cepheids had to be measured directly.

1

At first, this proved possible for just a handful of Cepheids; but crucially there were stars in the Hyades Cluster, whose distance could be measured, that are similar to stars in clusters that contain Cepheids. These provided the calibration, so that the distance to any Cepheid could be worked out simply by measuring its period, calculating its intrinsic brightness, and comparing that with its apparent brightness. The technique only just worked, even as late as the 1980s, because there were only 18 Cepheids with reliably determined distances. However, data from the HIPPARCOS satellite, which became available

3

1. Henrietta Swan Leavitt, who discovered the value of Cepheid stars as distance indicators.

2. The small Magellanic Cloud lies 20,000 light years away from us and contains many Cepheid variables.

3. Everybody who lives in the Northern Hemisphere has seen at least one Cepheid variable – the bright star in this picture.

4. If we know how bright a star really is we can work out its distance from us by judging how faint it looks – just like street lights.

in the 1990s, provided direct distance measurements to some Cepheids, and improved the accuracy of the distance measurement for the Hyades Cluster, so that the Cepheid distance scale is more solidly based now than ever before.

Into the Universe

Other objects within the same galaxies (most notably supernova explosions) can now be calibrated for brightness against the Cepheid distance scale. Because supernovae are so bright, they can be used as distance indicators in galaxies far, far away.

4

TOPIC LINKS

1.1 Stepping Stones to the Universe
p. 24 Beyond the Milky Way

2.1 The Big Bang
p. 80 Beyond the Local Group

2.2 Cosmology for Beginners
p. 95 Einstein's geometry

1.2 STAR BIRTH

Stars are not born in isolation, but in nurseries that may contain thousands or even millions of stars forming together. No astronomer has been able to watch a single individual star being born, but because we see stars in all stages of the process, it is possible to work out how it happens. In the same way, an alien visitor to Earth could work out the life cycle of a human being without waiting to watch an individual person being born, living and dying, but by studying a large group of people where all stages of the life cycle are present.

Fortunately for astronomy, our Solar System is located in a densely populated part of the Galaxy, close to a stellar nursery where new stars are being born all the time. This is the Orion Nebula. The nebula is visible through a small telescope or binoculars as a faint patch of light just below the belt of the constellation Orion, the Hunter.

THE ORION NURSERY

The constellation of Orion, the Hunter, is one of the most familiar in the sky. If you look carefully just below the three stars that form Orion's belt, you can see a fuzzy patch of light, which is called the Orion Nebula (it is quite clearly visible with a decent pair of binoculars). This nebula is typical of the clouds of gas and dust that are scattered throughout the disc of the Milky Way Galaxy and are particularly associated with the spiral arms. Most of the known nebulae can only be identified with the aid of telescopes, but the Orion Nebula is visible to the naked eye because it is so close to us, a mere 1500 light years (about 460 parsecs) away.

Because it is so close, when large astronomical instruments, such as the Hubble Space Telescope, are turned on the nebula, they show its structure in spectacular detail, and reveal the presence of hundreds of newly formed, hot young stars. These light up the nebula from within and make the glow that we can see with our unaided eyes or binoculars.

Although the Orion Nebula looks spectacular because it is so close, it is important to appreciate that it is not special in any way; it is typical of the kind of stellar nursery in which new stars are being born in our Galaxy all the time. The Orion Nebula is about 20 light years (16 parsecs) across, and contains four big, bright stars forming a pattern of stars called the Trapezium and many hundreds of smaller stars more or less the same as the Sun was just after it was born.

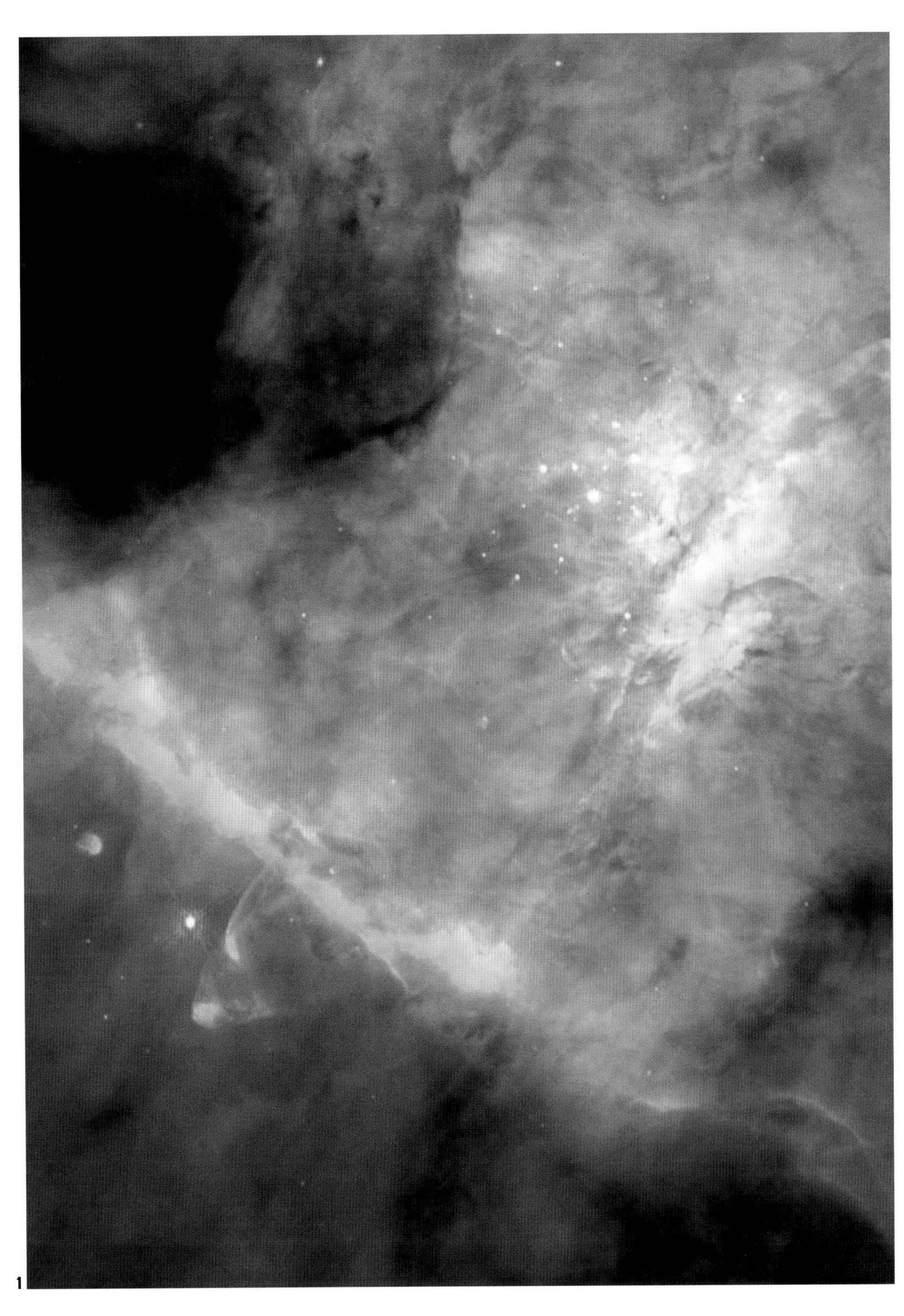

1. The great nebula in Orion is a region where new stars are forming today.

Turning up the Stellar Heat

Stars start to form in a gas cloud like the Orion Nebula because such clouds cannot be perfectly smooth and uniform. Some parts of the cloud are inevitably more dense than

1. The shape of these dark fingers of gas and dust in the Eagle nebula has been sculpted by radiation from young stars. Each finger is about one light year long.

2. Gas in the Orion nebula reflects the blue light of hot young stars.

others, and begin to attract material to themselves and collapse through the attraction of gravity. The more dense such a clump becomes, the stronger its gravitational influence becomes, and the more it attracts material from its surroundings.

As this proto-star collapses, it gets hot inside. This heat makes the young star begin to glow, and the heat clears a bubble of more or less empty space around it within the nebula. At first, the heat of the young star comes from the gravitational energy liberated during its collapse. But when the temperature at the heart of the star reaches a little more than 10 million degrees Celsius, the nuclei of hydrogen atoms can begin to fuse together to make new helium atoms. This can only happen if the proto-star has more than about one-tenth of the mass of our Sun. Blobs of gas with less mass than this cool off and settle down into objects rather like the giant planet Jupiter (which itself has 0.1 per cent as much mass as our Sun), only bigger. They become failed stars, which are sometimes referred to as 'brown dwarfs'.

Nebulae like the ones in Orion may contain enough gas and dust to make millions of stars like the Sun, but not all of this gas will be turned into stars, because as the first stars begin to form, the light and heat from the star-forming process blows the remaining material away. The material can easily be blown away like this because it is almost entirely made up of hydrogen and helium gas left over from the Big Bang when the Universe was born. In round terms, the material from which stars are made starts out as just under 75 per cent hydrogen and just under 25 per cent helium, which leaves no more than 1 per cent for everything else put together.

Stars like the Sun

For stars with enough mass, nuclear burning occurs when four hydrogen nuclei (protons) combine to make one helium nucleus in a multi-step process. Helium nuclei are so stable, and so important in the process of star formation (▷ p.53), that they are sometimes regarded as individual entities in their own right and called 'alpha particles'. Crucially, the mass of one alpha particle is a little bit less than the mass of four protons added together. The mass that is lost when an alpha particle is made in this way is turned into pure energy, in line with Albert Einstein's famous equation $E = mc^2$. The heat liberated as mass is converted into energy which provides a pressure that counters the inward pull of gravity and stops the star from contracting any more. It settles down as a stable star, burning steadily at the same temperature as long as its supply of nuclear fuel lasts. The stars we can see today in the Orion Nebula stabilized in this way about 300,000 years ago, very recently indeed by astronomical timescales.

Although stars usually form together in a nebula, they do not stay together throughout their lives. Like everything else in the disc of our Galaxy, the Orion Nebula and the stars it contains are orbiting around the centre. As the nebula disperses and the stars continue in their own orbits, they will each be moving at a slightly different speed, and although at first they will seem to move together, as a so-called 'open cluster' of stars, over hundreds of millions of years they will disperse into the disc of the Galaxy. The Sun formed some 4.5 billion years ago and it is now impossible to identify which of the hundreds of billions of stars in our Galaxy were born with it in the same nebular stellar nursery.

HEAT FROM GRAVITY

When a cloud of gas in space collapses under the pull of gravity (if you like, under its own weight) heat is released. But heat is a form of energy, and you can't get energy out of nothing. The heat of a young star is actually gravitational energy that has been converted into heat as the cloud collapses. On Earth, if you drop something from a tall building or into water, an object moves faster and faster until it hits a surface. As it falls, gravitational energy is being converted into energy of motion (kinetic energy). When it hits the surface, the kinetic energy is turned into heat energy, making the atoms and molecules in the object jostle around faster. Heat is a form of kinetic energy, related to the speed that atoms and molecules move.

In a collapsing gas cloud, all the atoms and molecules are trying to fall towards the centre of the cloud, so they move faster and faster until they hit something. The only 'something' around for them to hit is each other. So the atoms and molecules bash together more and more violently as the cloud collapses. Their kinetic energy increases because they are moving faster and the result is that the cloud gets hotter.

The four stars in the Trapezium constellation, are separated from each other by only 0.1 light years, just 2 per cent of the distance from the Sun to its nearest stellar neighbour.

DUSTY BEGINNINGS

It isn't just stars that form in nebulae such as the Orion Nebula – many of the stars forming in the Milky Way today probably have planets forming around them as they collapse and nuclear burning begins in their hearts. This is thanks to that 1 per cent or so of stuff in these nebulae that is not hydrogen and helium. This material is mainly other gases (such as carbon monoxide), but also very fine grains of dust, similar in size to the particles in cigarette smoke, made of carbon, silicon and other substances. All of this material has been produced inside previous generations of stars (▷ p. 67) and has been scattered into space as those stars age and die.

Apart from the presence of the Sun itself, the most important feature of the Solar System is that all of the material in it (with a few tiny exceptions) is orbiting in the same direction. The Sun itself rotates on its axis (once every 25.4 days or so), the planets (including the Earth) orbit around the Sun in the same direction, the various moons in the Solar System orbit their parent planets in this same direction, and so on.

What this tells us is that the cloud of gas and dust from which the Solar System formed was itself rotating in this direction –

1. Young stars in an association known as the Antares-Rho Ophiuchi complex.

no real surprise, since it is very unlikely that any cloud of gas will just hang around in space without rotating. When the cloud began to collapse, its rotation would have got faster, in exactly the same way that spinning ice skaters rotate faster by pulling their arms in to their sides. This is a feature of something called angular momentum, which all rotating objects have. The amount of angular momentum depends on the mass of the object, how this is distributed, and how fast the object is rotating. If most of the mass is far out from the centre, it will have more angular momentum than if it is concentrated near the centre; and if the object is rotating fast, it will have more than if it rotates slowly. So when a spinning object shrinks in size, it spins faster to keep the same angular momentum.

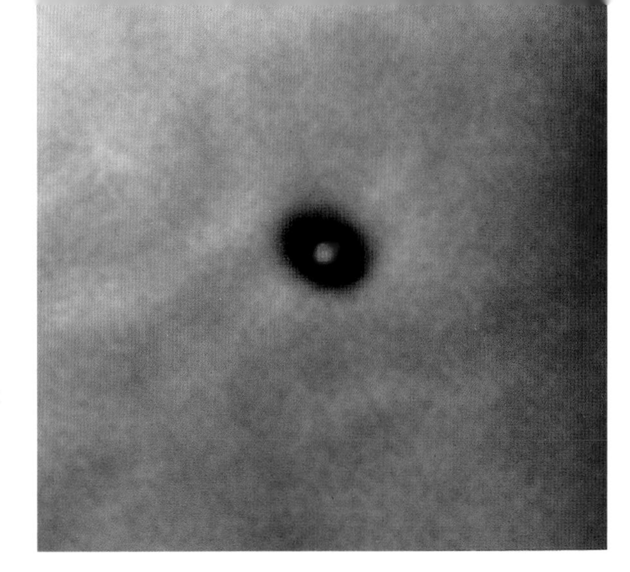

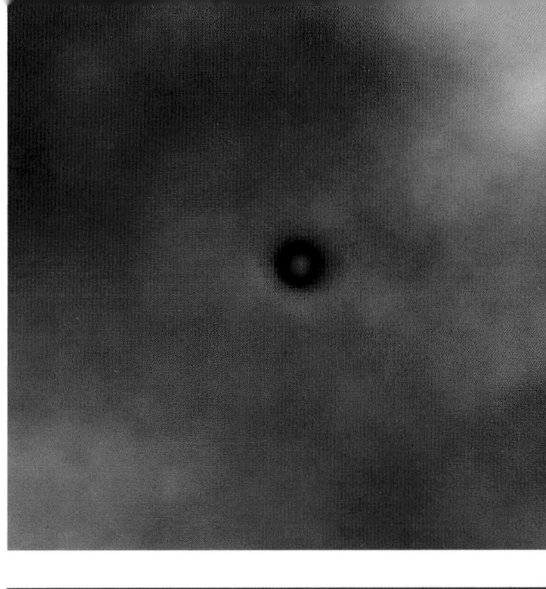

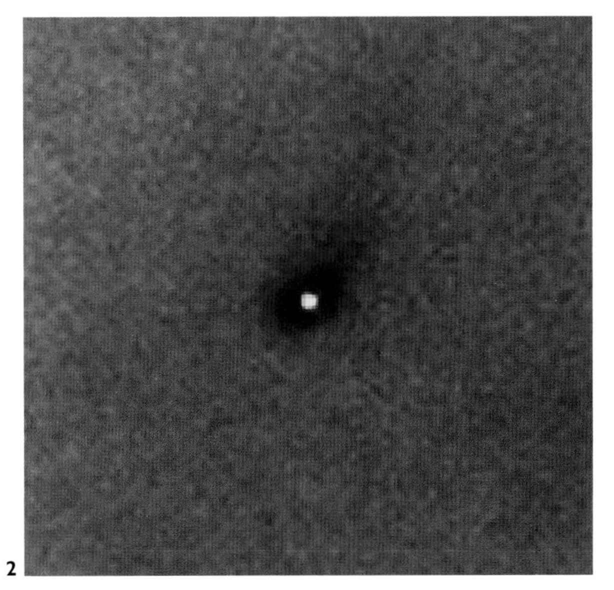

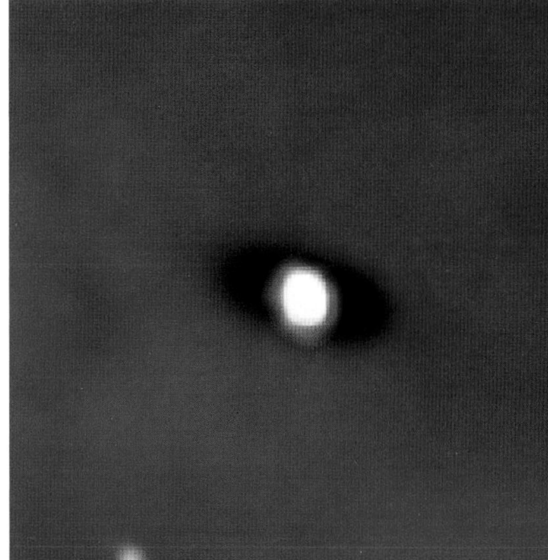

In a Spin

The collapsing cloud that formed the Sun and planets got rid of some of its angular momentum by throwing material off into space, like water flung from a rotating garden sprinkler. But even so, there was too much angular momentum left to allow all the dust associated with the young star to fall into it. As the star began to shine at the heart of the collapsing cloud, the leftover dust settled into a disc around it, a disc containing only a little of the mass of the system but, thanks to its position far out from the centre, lots of the angular momentum.

From Dust to Pebbles

The planets began to form in the dusty cloud even before it had settled into a disc. The tiny grains of dust in the cloud collided with each other more and more frequently as the cloud collapsed, because the volume of the cloud ▷▷

2. Disks of dusty material (protoplanetary disks) photographed around young stars in the Orion nebula.

3. As stars shrink they spin faster, like an ice-skater pulling in her arms.

FORMATION OF THE SOLAR SYSTEM

2. As the star forms, matter settles into a disc around it
3. Matter in the disc begins to clump together
4. Rocks form around the star
7. Planets like the Earth coalesce out of the rocky material

1

1. Family portrait representing the planets in our Solar System (excepting Pluto).

2. The Solar System also contains many smaller pieces of material – cosmic rubble such as comets.

was getting smaller, and there was less space between the grains. These very gentle collisions allowed the grains to stick together to make fluffy particles a few millimetres across, and these supergrains then collided with other supergrains to make bigger particles. Eventually, so many grains stuck together that objects the size of pebbles, then bigger lumps of rock formed, and gravity began to play its part. Once the primordial rocks were big enough, gravity would have pulled them together, and then the biggest rocks, by now kilometres across, would begin to dominate, attracting smaller lumps of rubble by gravity and growing bigger still.

As the lumps of rock grew, they collided with other lumps of rock. But because all the proto-planets, as they now were, were moving in the same direction around the Sun, these collisions would be relatively gentle, allowing the rocky lumps to join together, rather than be blasted back into dust. The larger lumps in the dusty disc grew to become planets, sweeping up most of the cosmic junk that still littered the disc. There is a reminder of those days still present in our Solar System. Between the orbits of the planets Mars and Jupiter, there is a band of rubble called the asteroid belt, populated by objects thought to be leftover material from the era of planet formation. There are estimated to be more than a million rocks in the asteroid belt bigger than 1 km across, with countless smaller lumps. The biggest, Ceres, is 933 km in diameter. But the total mass of all those rocks put together is only 15 per cent of the mass of our Moon.

Two Kinds of Planet

This picture of how planets formed neatly explains why there are two main kinds of planet in our Solar System. Close to the Sun, its heat would have driven away volatile material, leaving solid balls of rock with only thin atmospheres, such as Mercury, Venus, Earth and Mars. Further out from the Sun, it would have been cool enough for the rocky cores of planets to hold onto large amounts of gas, forming the giant planets Jupiter, Saturn, Uranus and Neptune. And further out still, there was scope for frozen balls of ice to remain in orbit around the Sun and that's where we find the ice planet Pluto and the comets that occasionally fall into our part of the Solar System from the outer fringes.

Recently, though, this simple picture has been challenged by the discovery of Jupiter-like

2

DUSTY DISCS

The astronomical explanation of how planetary systems form is no longer pure speculation, but is based on observations made in the 1990s of dusty discs of material around many young stars. The best studied of these discs is associated with a star called Beta Pictoris (right), and covers a span of at least 1000 AU. This is huge compared to the Solar System (the outermost giant planet, Neptune, is only 30 AU from the Sun), and shows that the disc is in the early stages of settling down, with some material still being ejected and blown away into space.

The Beta Pictoris system is thought to be only about 200 million years old, and the mass of material in the disc today is about one and a half times the mass of the Sun. Most of this will be lost as the system settles down. The inner part of the disc, comparable in size to our Solar System, is warped and distorted, possibly by the gravitational influence of planets forming within it.

The Hubble Space Telescope discovered hundreds of discs like this around young stars. This discovery is of key importance in our understanding of how planets form, and the fact that there are dusty discs around so many young stars suggests that planet formation is a common process. Further investigations of these objects will tell us what the Sun was like when the planets formed.

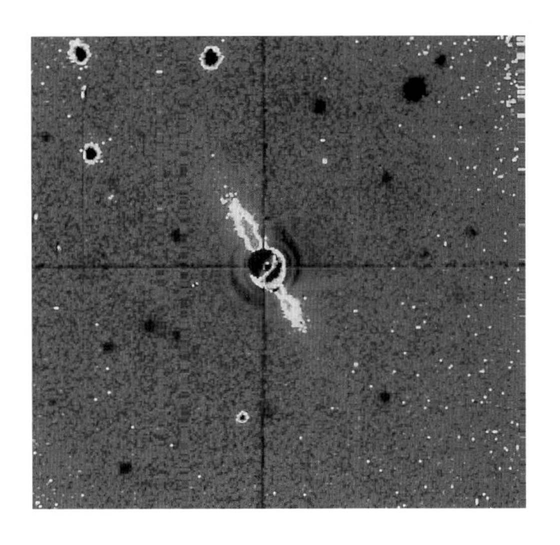

giant planets in Mercury-like orbits around other stars. Astronomers do not yet understand exactly how systems like this can form, although the discoveries are further evidence that solar systems are common in the Universe.

Whatever the details, the important point is that planet formation is a natural consequence of the way a dusty cloud of material collapses to form a single star. However this doesn't necessarily mean that all stars have planets. Multiple star systems have no problem getting rid of angular momentum, because it is stored in their orbital motion around each other, and it is unlikely that there are stable planetary orbits in these more complicated systems. Rather more than half of all stars seem to occur in systems with at least one other stellar

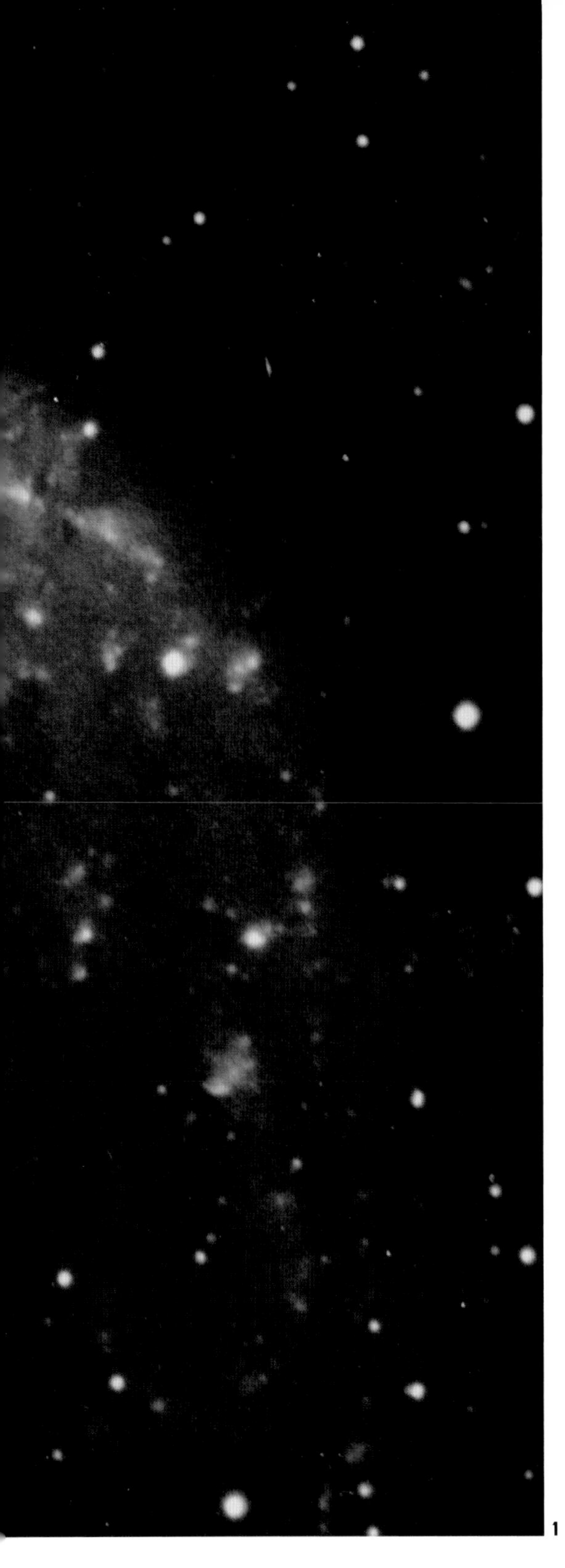

THE COSMIC SQUEEZE

Astronomers have a good understanding of how stars and planets form when a cloud of gas and dust in space begins to collapse. But what is it that triggers the collapse in the first place? In the Milky Way (and other disc galaxies) today, stars continue to be born because of the activity associated with the presence of other stars in the galactic disc itself.

The Orion Nebula is part of a much bigger cloud of gas, called a 'giant molecular cloud', which covers almost all of the constellation Orion on the night sky. Such giant molecular clouds are held together by gravity and can be regarded as single entities, the most massive single entities in our Galaxy, with a mass up to 10 million times the mass of the Sun and diameter of between 46 and 77 parsecs (150 and 250 light years). If a star explodes to create a supernova on one side of such a cloud, it sends out shock waves in the form of ripples moving through the cloud. These shock waves sweep up material in front of them (almost literally in the way that a broom sweeps dust into a heap), and this compresses some of the gas in the cloud enough to start it collapsing to form new stars with a whole range of masses. The biggest blobs of gas in these compressed regions collapse very quickly and form massive stars – tens of times as massive as our Sun which run through their life cycles very rapidly (in less than a million years) and explode in turn, sending more ripples out across the molecular cloud. In this way, a wave of star formation can travel right across a giant molecular cloud over a span of 10 or 20 million years.

companion (some have two or three). But even if just under half of the stars in the disc of the Milky Way are single systems like our Sun, that still leaves scope for well over 100 billion solar systems to have formed in our Galaxy in the way just described.

How Galaxies Make Stars

This is still not the complete picture, because the molecular clouds themselves are associated with the spiral arms of a galaxy like our own. They have come into existence because the thin

1. Planetary systems like our Solar System seem to be a natural part of spiral galaxies like the Milky Way.

gas between the stars has itself been squeezed on its journey around the disc of the Galaxy. Although the pattern made by the spiral arms twining outwards from the nucleus of a disc galaxy looks superficially rather like the pattern of cream stirred into a cup of coffee, there is one important difference. The spiral pattern made by the white cream against the black coffee quickly dissolves into a smooth brown colour as it mixes in. In the same way, the spiral pattern in a disc galaxy ought to get smeared out as the individual stars move on in their own orbits, at their own speeds around the centre. This should happen within about a billion years, a blink of an eye in the lifetime of a galaxy. But the spiral arms are not dissolved by rotation because they are constantly being renewed.

The distinctive pattern that shows up so clearly on photographs of spiral galaxies is caused by the presence of hot young stars along the edges of the spiral arms. These hot young stars are being born there because new clouds of gas and dust are constantly moving through the spiral pattern and being squeezed by a shock wave there. It is the shock wave that is the more or less permanent feature, like a sonic boom spiralling around the Galaxy, with the gas and dust just passing through. A useful way to picture what is going on is to think of a crowded motorway, where a slow-moving, large vehicle causes a kind of mobile traffic jam. Vehicles pile up behind the obstruction, work their way slowly past it, then speed up on the other side. The traffic jam moves along the road at a steady speed, but the individual cars that make up the jam are constantly changing as new cars approach from behind and others escape from the front of the jam. In the case of the spiral arms of our Galaxy, the shock wave itself is travelling around the Galaxy at a speed of about 30 km per second, while the stars and clouds of gas and dust are overtaking the spiral wave at a speed of about 250 km per second, passing through it but getting squeezed in the process. Clouds of gas and dust pile up in the cosmic traffic jam along the inside curve of the spiral arm, and are squeezed there, triggering bursts of star formation like the one going on in the Orion molecular cloud.

How Stars Make Stars

The whole process is self-sustaining. The shock waves are maintained by the stellar explosions going on along the edge of the spiral arm, and the stellar explosions are caused by the squeeze given to clouds of gas and dust by the shock waves. Although giant molecular clouds are constantly forming stars in this process, the explosions of other stars constantly feed raw material back into interstellar space to make new molecular clouds, the raw material for later generations of stars. Each year, just a few solar masses of material are being recycled in this way in the Milky Way Galaxy, but over a few billion years that adds up to a lot of new stars.

At any one time the amount of matter orbiting around the disc of the Galaxy in the form of giant molecular clouds is about 3 billion times as much as there is in our Sun. This is equivalent to about 15 per cent of the total mass of stars in the disc itself. It takes surprisingly few big stellar explosions to keep the recycling process going. There are only two or three supernova explosions in our Galaxy every century, and there hasn't been one close enough to be studied by human observers for nearly 400 years (the closest occurred in 1987 in a small nearby galaxy, the Large Magellanic Cloud). Think of this in terms of the lifetime of a galaxy. Even at a rate of only two per century, there will be 20,000 supernova explosions every million years, and several hundred million have occurred since the Sun was born 4.5 billion years ago.

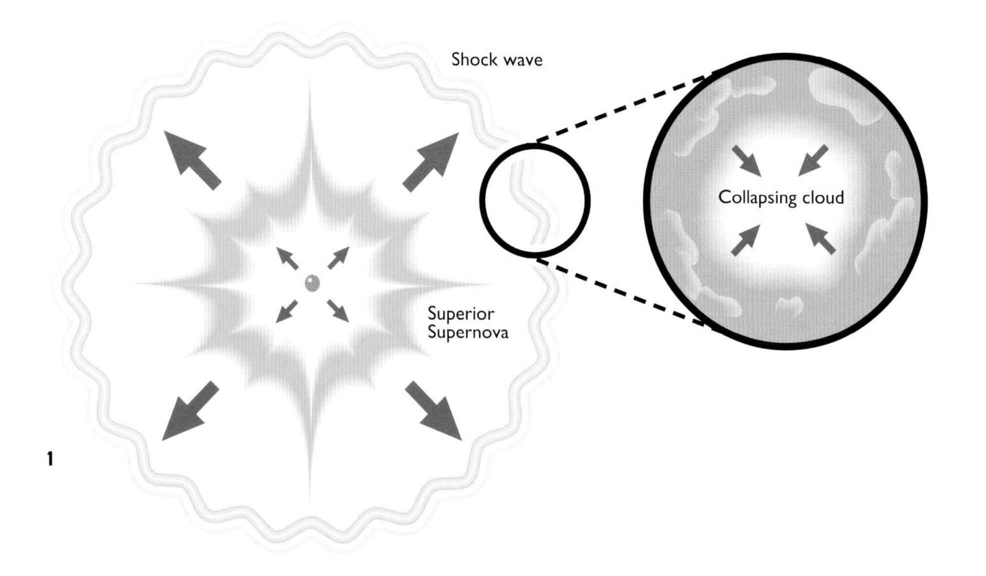

1. The expanding shockwave from an exploding star (main picture) squeezes clouds of gas and encourages new stars to form (inset).

2. (opposite) The large Magellanic Cloud is a near-neighbour of our galaxy. It is about 160,000 light years away, and 30,000 light years across.

1.2

OUR COSMIC NEIGHBOURHOOD

The Sun and its family of planets lie just inside one of the spiral arms of our Galaxy, passing through it on their endless journey around the Milky Way. This spiral arm is sometimes called the Orion Arm, after the distinctive constellation of Orion. But because the bright stars that form the traditional constellations lie in the same spiral arm, this name is slightly misleading and many astronomers prefer to call our region of the Galaxy simply the 'Local Arm'.

The View from Inside

From our viewpoint within the Local Arm, astronomers see a congregation of stars, nebulae and dark clouds in two opposite directions on the sky. The Local Arm comes outwards from the inner part of the Galaxy from the direction of the constellation Cygnus, as seen from Earth, and carries on past us in the direction of the constellation Vela. The Orion Nebula and its nursery of stars are slightly offset from this main curve of the Local Arm and lie about 46 parsecs (1500 light years) away from us. The brightest star in the sky, Sirius, is part of the constellation Canis Major, and is also known as Alpha Canis Major, meaning that it is the brightest star in that constellation. But Sirius is not intrinsically very bright; it only looks bright because it is so close to us, 2.5 parsecs (8.7 light years) away, the seventh closest star to the Sun.

1

Canis Major seems to be part of a huge, faint nebula, with a diameter of 100 light years, which is sometimes called the Seagull Nebula after a fancied resemblance to a bird. But appearances can be deceptive, because of the way objects at different distances are superimposed on the sky. In fact, the Seagull Nebula lies far beyond Sirius, roughly a kiloparsec (3000 light years) away. The nebula is an expanding shell of material produced in a supernova explosion about half a million years ago and is associated with a group of young stars (about 300,000 years old) whose birth was triggered by the shock wave from the supernova.

2

Local Nurseries

Between us and the Seagull Nebula there are three regions of active star formation, the closest being the Orion Molecular Cloud. All three regions seem to be linked by a thin streamer of gas, only 6 parsecs (20 light years) across but more than 300 parsecs (1000 light years) long.

Apart from the regions of star birth (and the stars themselves), there is another key component of the story of stellar life and death to be seen in our neighbourhood. The Helix Nebula, lying in the direction of the constellation Aquarius, but about 140 parsecs (450 light years) away, is the closest example of a so-called 'planetary nebula'.

3

1. The constellation Canis Major showing Sirius, the brightest star in the night sky.

2. The Helix nebula, also known as NGC 7293.

3. The constellation Orion.

Overleaf. The Trifid Nebula in the constellation Sagittarius. The pink part is a cloud of hydrogen gas, the blue area reflects the light of bright stars. Dark lanes of dust obscure emitted light.

Planetary nebulae get their name because, when viewed through a small telescope, they look like discs on the sky, just as planets do. But in fact they are huge, expanding balls of gas that have been ejected by a star late in its life. The Helix Nebula is so big that even at a distance of 450 light years it covers half of the apparent diameter of our Moon on the sky. It is a glowing shell of material (nearly a light year across) ejected from the central star about 10,000 years ago.

Every part of the cycle of stellar birth, life and death is represented in our immediate cosmic neighbourhood. This is why astronomers have been able to piece together the life story of a star. If we had not been in a spiral arm, astronomers might not have been able to do this.

The gas in any part of the disc of our Galaxy is exposed to the squeezing effect of a supernova at least once every few million years, on average, while it takes the Sun more than 200 million years to orbit around the Galaxy just once. However, the average is a little misleading, because most of the explosions, and most of the squeezing, happen near spiral arms, with long quiet periods during the journey of a star around the Galaxy.

The Sun has orbited right around the Milky Way Galaxy about 20 times since the Solar System formed.

TOPIC LINKS

1.1 Stepping Stones to the Universe
p. 21 An island in space

1.2 Star Birth
p. 31 The Orion Nursery

1.3 Stellar Evolution
p. 51 Lifetimes of the Stars
p. 57 Red Giants and White Dwarfs

1.3 STELLAR EVOLUTION

Astronomers use the term 'evolution' to refer to the life cycle of a star: how an individual star changes as it goes through its life cycle. This is different from the way the term is used in biology, where it refers to the way a population of individuals changes from generation to generation.

Once nuclear burning begins to occur in their hearts, all stars stabilize and shine more or less steadily as long as their nuclear fuel lasts. During this stable phase of their lives, stars range in size from one-tenth to 10 times of the diameter of our Sun. The size and the external appearance of such a stable star (its brightness and colour), and the amount of time that it can spend burning nuclear fuel steadily in this way, depend on just one thing: its mass. Bigger stars burn their fuel more quickly to hold themselves up; smaller stars can sustain themselves with less rapid consumption of fuel.

THE MAIN SEQUENCE

The process which generates the heat that provides the pressure holding a star up and prevents it from collapsing under its own weight is the conversion of hydrogen into helium. This is a multi-step process (actually, two multi-step processes – ▷ p. 58) in which a little of the original mass is converted into pure energy. The higher the mass of a star, the more weight it has to support. The only way that the star can do this is to burn its nuclear fuel more rapidly, generating a higher pressure in its core. This means that it will use up its fuel more quickly. It also means that more heat is escaping from the surface of the star. So a more massive star is both shorter-lived and brighter than a less massive star.

If two objects are the same size, but have different temperatures, they will have different colours. For example, a red-hot lump of iron is cooler than a white-hot lump of iron. If all stars were the same size, the same simple rule would apply to them, and the colour of a star would immediately tell you its temperature and therefore reveal its mass. But the astronomer's life is not quite that simple.

If one star is bigger than another, it has a larger surface area from which heat can escape. So the same amount of heat can

1

1. The colours of stars depend on how hot and bright they are, which helps astronomers to work out how far away they are.

escape from the whole surface of the larger star every second, even though each square metre of its surface has a lower temperature than each square metre of the surface of the smaller star; there are more square metres for the heat to escape from. In principle, if the size difference is big enough, a red star can be radiating just as much heat as a white star. And it is the *rate* at which heat is escaping overall which tells us how much heat is being generated in the core of the star, and therefore how massive it is.

The Key to Astrophysics

When, in the twentieth century, astronomers began to study the relationship between the colour of a star and its absolute brightness (the brightness it actually has close up, not its apparent brightness on the sky), they found that many stars obey a simple rule, which shows up clearly when absolute brightness (which astronomers refer to as 'absolute magnitude') is plotted against colour in a kind of graph known as a colour–magnitude diagram. The colour–magnitude diagram is also known as the Hertzprung–Russell (or HR) diagram, after the two astronomers who discovered the relationship.

The HR diagram is plotted in such a way that the brightness of a star (its absolute magnitude) increases as you go up the page, and the temperature decreases as you go from left to right across the page. So stars at the bottom right of the diagram are faint, cool and red (with surface temperatures below about 3500 Kelvin) while stars in the top left are bright, hot and blue-white (with surface temperatures above about 25,000 K). Most visible stars lie on a band in the diagram (opposite page) running from the top left to the bottom right, and this band is called the 'main sequence'. All stars that, like our Sun, get their energy by converting hydrogen into helium in their hearts, lie on the main sequence.

Large Stars are Hot

From other evidence relating to the masses of stars (in particular, the way that stars in binary systems orbit around each other), astronomers have been able to measure

ABSOLUTE TEMPERATURE

Like other scientists, astronomers prefer to measure temperatures in terms of the absolute, or Kelvin, scale. Unlike the temperature scales in everyday use (the Celsius and Fahrenheit scales), the Kelvin scale is not based on any arbitrary choice of a zero point, such as the freezing point of water, but on the absolute zero of temperature, the lowest temperature that can ever exist. This temperature was calculated from thermodynamic principles by the British physicist Lord Kelvin (left) in the nineteenth century and corresponds to –273.16 degrees on the Celsius scale. The size of the units on the Kelvin scale has been chosen to match the size of the degrees on the Celsius scale, so the freezing point of water is 273.16 kelvin, or 273.16 K (on the Kelvin scale, the degree symbol is never used). Astronomical temperatures are often so high that it doesn't make very much difference which units you use. The temperature at the surface of the Sun, for example, is 5800 K, which corresponds to 5527 degrees Celsius, a difference of less than 5 per cent. The choice of units is even less important when we describe the temperature at the heart of the Sun – 15 million degrees Celsius or kelvin means much the same. However, it is crucially important to be sure which units are being used when describing the temperatures of cold objects in space, where the temperature can fall to within a few kelvin of absolute zero.

enough stellar masses directly to understand what is going on in the main sequence. A small star, which does not have to burn fuel very rapidly in order to hold itself up, sits at the bottom of the main sequence; while a large star that has to burn fuel rapidly in order to avoid collapse sits at the top of the main sequence. The Sun is a middle-sized star, generating a moderate amount of heat and having a yellow-orange colour; it sits near, but slightly below, the middle of the main sequence.

Although most visible stars lie on the main sequence of the HR diagram, there are exceptions. In particular, there are a few visible stars in the bottom left half of the diagram, and some in the top right half of the diagram. The ones at the bottom left are small, faint and hot, while the ones at the upper right are large, bright and cool. They represent stars which are in the later stages of their evolution. They have left the main sequence.

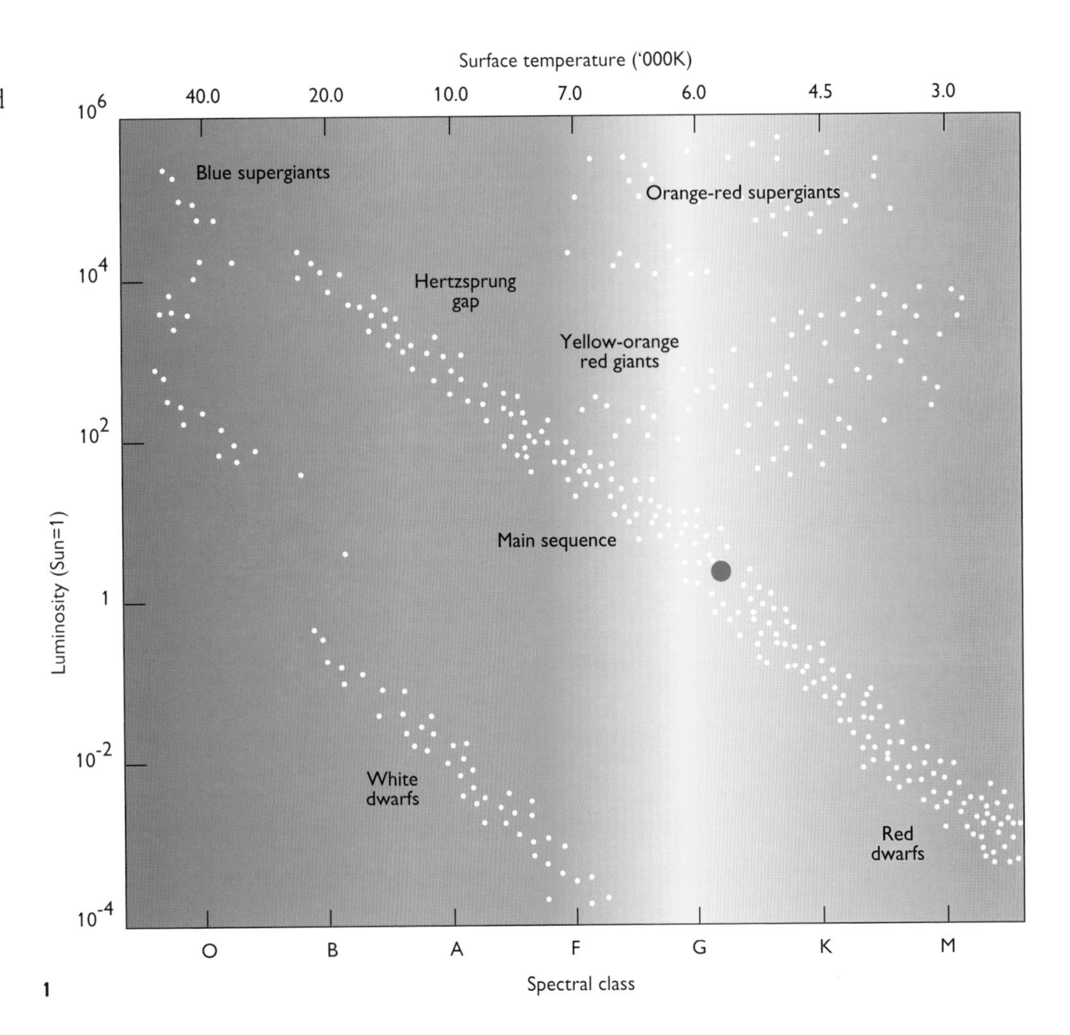

1

Lifetimes of the Stars

The amount of time a star can spend on the main sequence depends on its mass, which also determines its position on the main sequence. So the bright stars at the top of the main sequence live fast and die young, like some rock stars. The Sun itself will be able to spend a total of about 10 billion years steadily burning hydrogen in its core as a main sequence star, and is now just under half way through its main sequence lifetime. A cool star with a mass about one-tenth of that of the Sun can sit at the bottom of the main sequence quietly burning its fuel for hundreds of billions of years. But a star with five times as much mass as the Sun only has a main sequence lifetime of some 70 million years, and those with 25 times the mass of the Sun have main sequence lifetimes of only

2

1. Schematic representation of the HR diagram. The position corresponding to the Sun is marked with a dot.

2. Betelgeuse is a massive red, supergiant star 17,000 times brighter than the sun and with a diameter 300 times bigger than that of the sun.

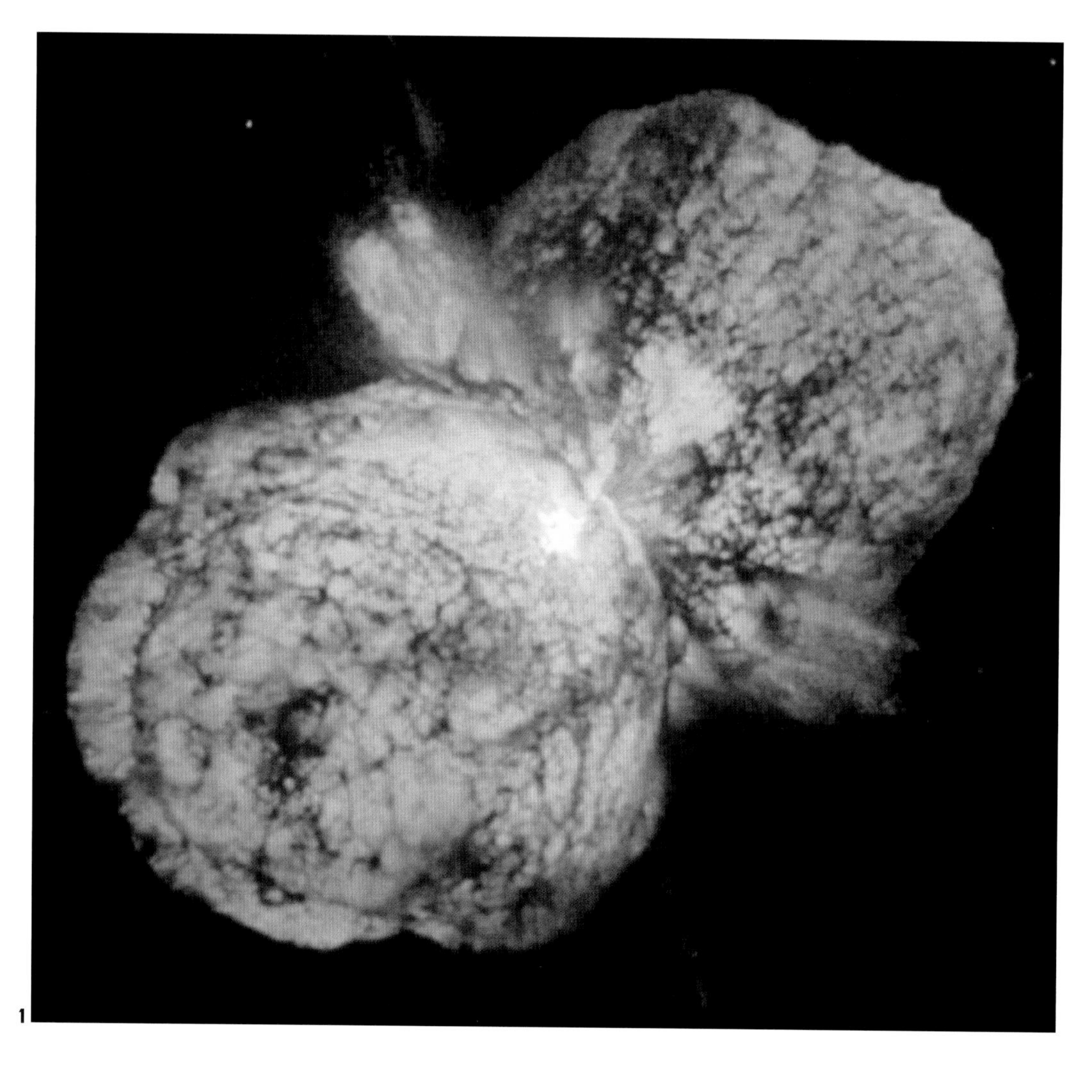
1

3 million years. The most massive stars on the main sequence have about 50 times the mass of our Sun, and are perhaps 20 times its diameter, but have lifetimes of less than a million years. Anything much bigger than this that tries to become a star generates so much heat in its core as it collapses that it blows itself apart.

Because of the way bigger stars have shorter lifetimes, when a group of stars of the same age but with different masses is plotted on an HR diagram the main sequence will be incomplete. This can be done for stars in globular clusters, which are groups of thousands (sometimes millions) of stars that were born together from a single cloud of material long ago. In such a cluster, the more massive stars have already left the main sequence, so the top left of the main sequence is truncated. The exact point where the main sequence ends tells you the masses of the oldest stars still on the main sequence in the cluster today, which (since they must be just about to leave the main sequence themselves) reveals their age, and that tells you how old the cluster is.

Brightness and Distance

The HR diagram can also be used to measure distances to clusters of stars, because the position of the main sequence on the diagram is related to the absolute brightnesses of the stars. The further away the cluster is, the lower down on the diagram the main sequence will seem to be, because the stars will look fainter. Astronomers can work out how far it is to the cluster by calculating how much brighter the stars would have to be to lie on the standard main sequence. All of this places the HR diagram and the main sequence among the most important sources of information about the stars.

☆ Although the core of the Sun has 12 times the density of lead, its outer layers are so tenuous that the average density of the whole Sun is only 1.4 times that of water.

COOKING THE ELEMENTS

All of the elements in the Universe today, except for hydrogen, helium and a very small amount of light elements such as lithium and deuterium, have been manufactured inside stars and spread through space late in the lifetimes of those stars, either as clouds of gas puffed away gently from the surface of the star or in great stellar explosions. The raw material for the manufacture of those elements is the hydrogen and helium which were produced in the Big Bang when the Universe was born. Some of that hydrogen is being converted into helium today inside main sequence stars, and more was converted into helium in the past. But by far the bulk of the 25 per cent helium that contributes to the composition of a star like the Sun is primordial stuff left over from the Big Bang. All the heavier elements, such as oxygen, carbon, nitrogen, iron, and all the elements in your body, are literally star dust created from this primordial matter.

The fundamental component of an element is, of course, the atom. All atoms of a particular element (such as oxygen) are identical to one another and are made up of a core (the nucleus) containing particles called protons and others called neutrons, surrounded by a cloud of electrons (one electron in the cloud for every proton in the nucleus). It is the number of protons in the nucleus which determines the nature of the

2

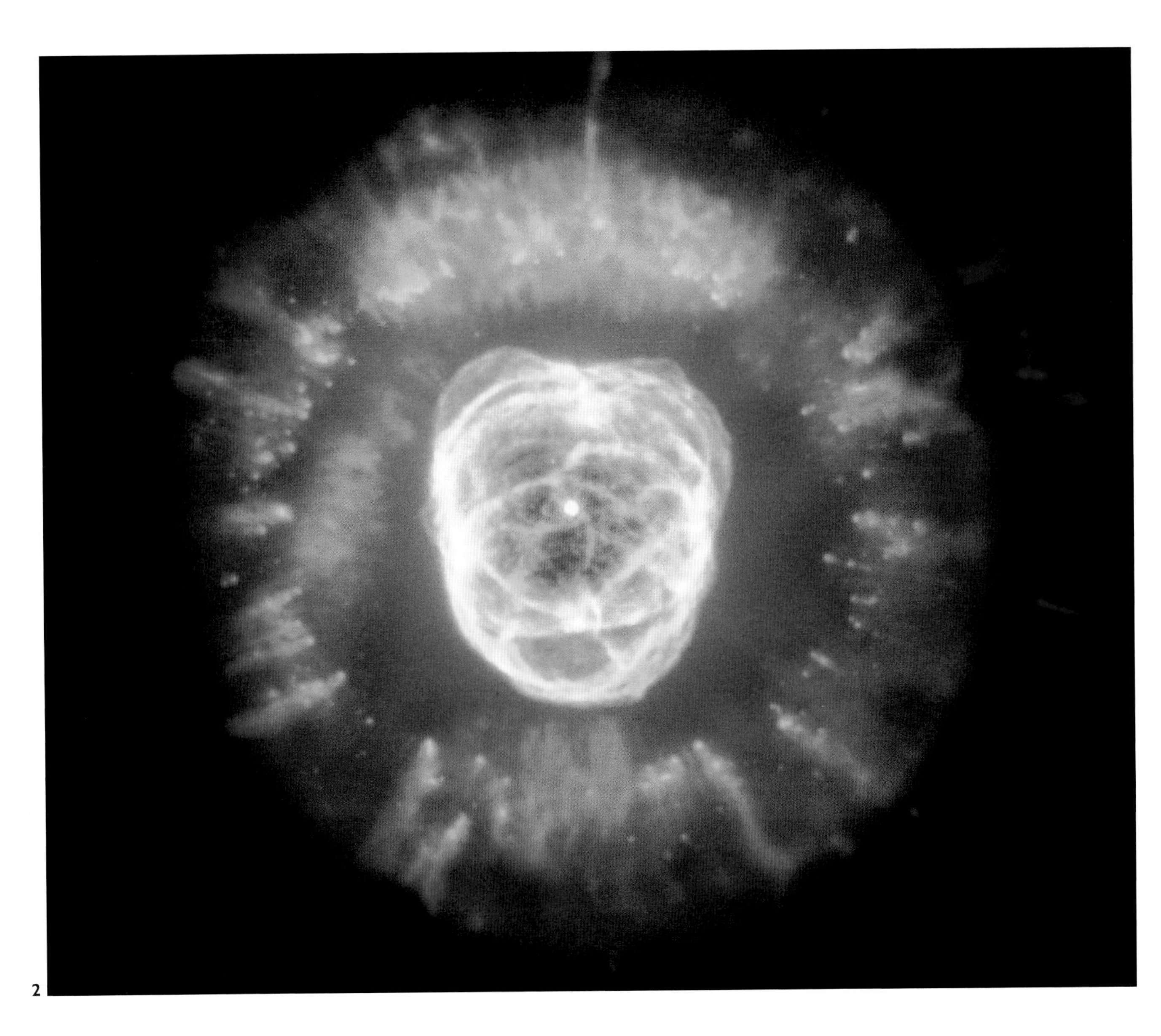

1. Eta Carinae is a star about 100 times as massive as our sun, blowing material away into space.

2. The Eskimo nebula is one of the prettiest examples of a planetary nebula.

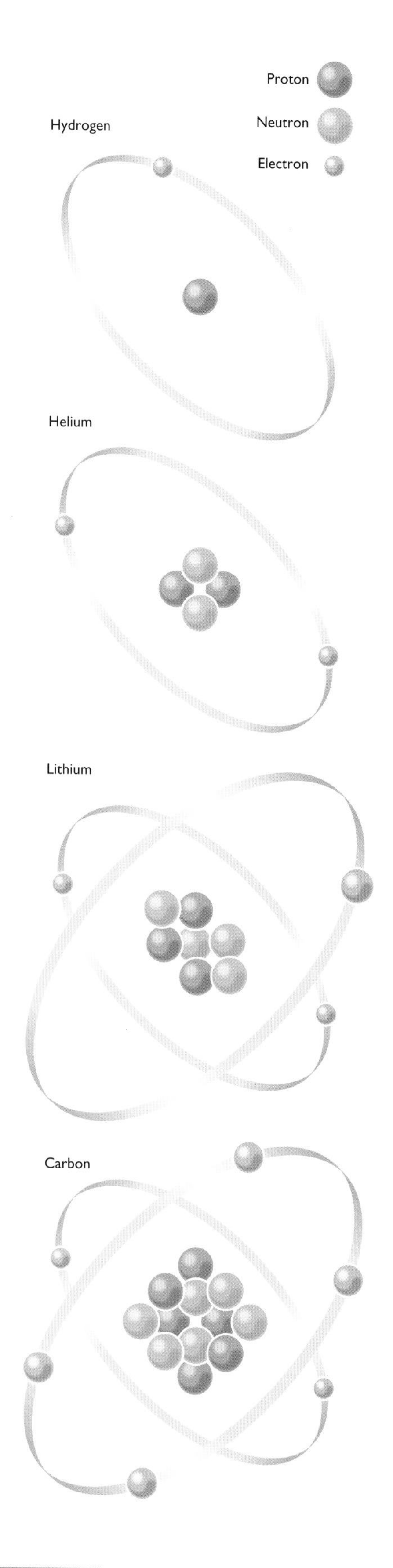

1

atom, whether it is an atom of lead, or sulphur, or whatever. The simplest atom of all, hydrogen, has a nucleus consisting of a single proton.

The Stellar Pressure Cooker

Under the extreme conditions that exist in the hearts of stars (temperatures in excess of 15 million K and densities greater than 10 times the density of lead on Earth), the electrons are stripped away from their atoms, and the nuclei are free to move about through a sea of free electrons and collide with one another. Very occasionally, the nuclei interact to make new nuclei, making new elements in the process known as stellar nucleosynthesis.

Making helium out of hydrogen is the first step in nucleosynthesis. Although this mostly happened in the Big Bang, the stellar process is still important in providing the energy that keeps main sequence stars shining. The story of how elements are cooked inside stars really begins later in a star's life, when conditions become extreme enough for helium nuclei to combine with one another to make nuclei of carbon.

The problem with nucleosynthesis is that it is difficult to make nuclei stick together. All nuclei carry positive electric charge, because protons are positively charged and neutrons have no charge at all. If the nuclei can come close enough together (in effect if they touch) then they can interact and (sometimes) fuse to make new nuclei, held together by a strong nuclear force which overcomes the electrical repulsion but which has only a very short range. Slow-moving nuclei are repelled from one another electrically before they can interact in this way; only fast nuclei (which means hot nuclei) can collide powerfully enough to overcome the electrical repulsion. The more protons there are in the nucleus, the stronger the repulsion is and the hotter the nuclei must be before they can interact.

The extra heat required to initiate stellar nucleosynthesis is provided, paradoxical though it may seem, when the core of a main sequence star has used up all of its hydrogen fuel. Because the core can no longer generate heat by converting hydrogen into helium, the temperature and pressure there fall, and the weight of the star pressing down on the core makes it collapse. But as the core begins to collapse, gravitational energy is released, making the core hotter again. It gets so hot that helium nuclei can move fast enough to stick together. Energy is released (through $E = mc^2$ ▷ p. 58) as they do so and the collapse is halted for as long as the supply of helium lasts.

Cooking up Carbon

Each helium nucleus involved in this process contains two protons and two neutrons (and is known, for obvious reasons, as helium-4). (There is another variety of helium containing two protons and a single neutron (helium-3), but it is not important here.) You might think that the natural result of an interaction between two helium-4 nuclei would be a nucleus containing four protons and four neutrons (beryllium-8). But it turns out that this nucleus is very unstable and breaks apart within 10 millionths of a billionth of a second after it forms. The only way that helium-4 nuclei can combine to form a new stable nucleus is if three of them get together at very nearly the same time to make a nucleus of carbon-12. This step is made much easier by the fact that the carbon-12 nucleus has just the right energy to be stable. The three helium nuclei initially form what is called an 'excited' carbon-12 nucleus, which then gets rid of the excess energy by radiating it away, forming a stable nucleus. This is called the 'triple alpha process'.

Just as a helium-4 nucleus has slightly less

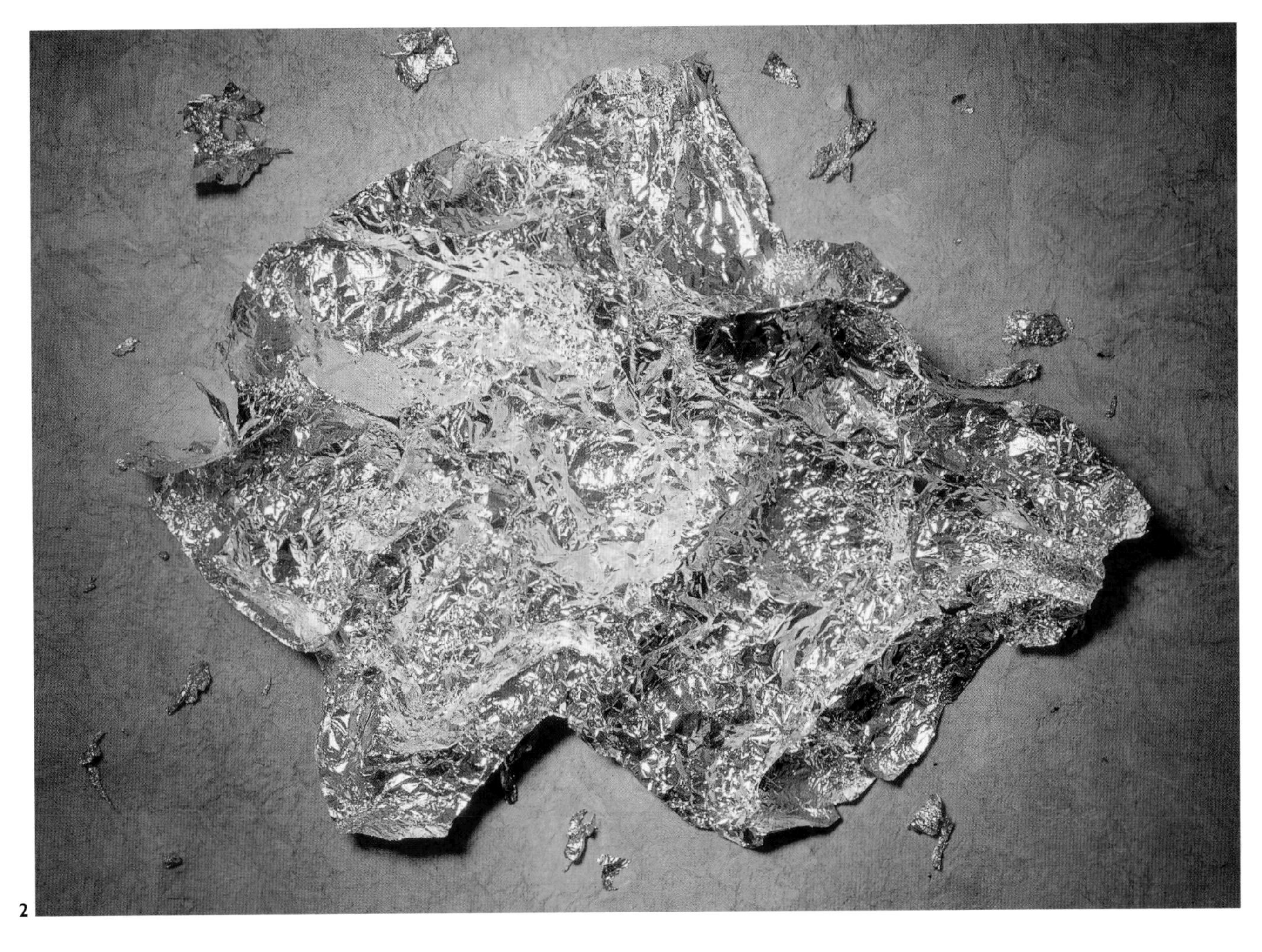

1. The atomic structure of some simple elements.

2. Heavy elements such as gold are only made in dying stars.

mass than the total masses of the two protons and two neutrons that make up the nucleus, so a carbon-12 nucleus has slightly less mass than the three helium-4 nuclei that went into it. This loss of mass is what provides the heat to keep a star going during this phase of its life, 'burning' helium into carbon.

Making Heavier Elements

Once carbon-12 has been formed, making the rest of the elements is downhill all the way. When the helium in the core of a star is beginning to be used up, once again the core shrinks a little, gravitational energy is released, the core gets hotter, and a new wave of fusion begins. This time, alpha particles fuse with carbon-12 nuclei to make oxygen-16. Further steps in the process (for stars which have enough weight to generate the required temperatures in their cores) manufacture the elements neon-20, magnesium-24 and silicon-28. Other elements are produced when the nuclei of these elements are involved in interactions in which they either absorb or emit protons, or in which neutrons are converted into protons, or vice versa.

But the end of the line comes when pairs of silicon-28 nuclei fuse together to make iron-56 and related elements, such as cobalt-56 and nickel-56. At every step, energy has

Every second, the Sun converts 5 million tonnes of matter into pure energy, in line with Einstein's famous equation $E = mc^2$.

1. The variable star Mira is shown here at its maximum brightness.

2. Alpha Hercules is an example of a red giant star.

been released by fusing lighter nuclei to make nuclei of heavier elements. But from the iron-group elements onwards, it takes energy to make the nuclei stick together to make nuclei of even heavier elements. There is no way to liberate energy by making iron-group elements fuse and any star that gets this far has no nuclear source of energy left to draw on.

There are, of course, elements heavier than iron in the Universe at large and on the Earth itself – like gold, lead, and uranium. They must have been made somewhere, and the story of their origin is one of the most dramatic in astronomy. But as far as ordinary stellar nucleosynthesis is concerned, iron and its associated family of elements marks the end of the line. Many stars (including the Sun itself) are not big enough for nucleosynthesis even to proceed that far, and will not be able to manufacture anything more interesting than carbon (indeed, the lightest stars will never get beyond the simple fusion of hydrogen nuclei to make helium). But all this activity in the core of a star has profound effects on its outer layers, and even a star like the Sun is destined to change its outward appearance drastically as its core adjusts to different phases of nuclear burning.

RED GIANTS AND WHITE DWARFS

A star stays on the main sequence of the HR diagram, with very little change in its outward appearance, as long as it is converting hydrogen nuclei into helium nuclei in its heart. But when hydrogen burning ends the appearance of the star changes dramatically.

At this point in its life, the core of the star shrinks and gets hotter, until helium burning begins. The effect of the extra heat coming out from the heart of the star is to make the outer layers (its atmosphere) swell up and expand. So the star gets a lot bigger overall, even though its core has shrunk. Because the star is much bigger, it has a much larger surface area from which the heat coming from the core can escape. This means that although there is more heat escaping overall, less heat is escaping across each square metre of the surface. So the surface gets cooler, even though the core gets hotter and the star gets brighter. A star like the Sun, which starts out yellow or orange on the main sequence will cool to a deep red colour as it expands in this way. It becomes a red giant, one of the bright but cool stars that lie in the upper right side of the HR diagram, above the main sequence (▷ p. 51).

Wandering Stars

If you could live long enough to watch a single star change as it aged, you would see it wander about in the HR diagram, moving upwards and to the right away from the main sequence at first, but zig-zagging about in the red giant part of the diagram as the conditions in its core changed and different nuclear energy sources came into play. During part of its time as a red giant, such a star may, if it has the right mass, experience rhythmic pulsations as its atmosphere swells and shrinks in response to the changes going on inside it. It is these pulsations that make some red giants show up as Cepheid variables (▷ pp. 28-9).

In another part of the HR diagram, red giants show a similar and almost equally useful kind of activity: they are known as RR Lyrae variables and are also good distance indicators, even though they are fainter than Cepheids.

A stable red giant star has a core in which helium is being burnt into carbon by the triple alpha process. This core is surrounded by a shell of material in which hydrogen is being converted into helium just as it is in a main sequence star, and this in turn is surrounded by a hugely distended atmosphere consisting ▷▷

THE FATE OF THE EARTH

Although a main sequence star like the Sun (right) stays roughly the same during its time on the main sequence, even minor changes in solar output can have a big effect on the Earth. Computer simulations of how stars work suggest that the Sun has actually got warmer, slightly, since it was formed some 5 billion years ago. So far, this has been compensated for on Earth by changes in the atmosphere of our planet, which have reduced the strength of the natural greenhouse effect. But if things continue in this way, the Earth may become uninhabitable in about a billion years time. Looking further ahead, there is about another 5–6 billion years before the Sun becomes a red giant. At that time, its diameter will increase to between 150 and 200 times its present size, and its brightness will increase to 2000 times its present value. By then, because the Sun will have lost about a quarter of its mass, loosening its gravitational grip on the planets, the Earth will have drifted out into a larger orbit. However, it will still be close enough to the Sun for the surface layers to be melted. As the Sun blows away its atmosphere to form a planetary nebula and settles down to become a white dwarf, all that will be left of the Earth will be a lump of solidified slag with no atmosphere, orbiting far out from the faint, dying star.

1.3

WHY STARS SHINE

Stars on the main sequence of the Hertzsprung–Russell diagram (▷ p. 51), like the Sun, shine because they are converting hydrogen into helium (specifically helium-4) in their cores, in the process known as 'nuclear fusion'. The nucleus of a hydrogen atom consists of a single proton. The nucleus of a helium atom is made up of two protons and two neutrons, held together by a short-range interaction called the 'strong nuclear force'. To make a helium nucleus from hydrogen, you start with four protons, persuade two of them to convert themselves into neutrons, and stick the resulting four particles together. This is not easy, but nature has found two ways to perform this trick – and both operate inside stars.

On the Chain Gang

The first process is called the 'proton–proton chain' (or p–p chain) and it is the main source of energy for the Sun and other stars in the bottom half of the main sequence of the Hertzsprung–Russell diagram. The process begins when two protons come close enough together to interact in spite of their mutual electrical repulsion. In the interaction, one of the protons spits out a 'positron' (a positively charged counterpart to an electron) and a very light particle called a 'neutrino', thus converting itself into a neutron.

With the positive charge of one of the original protons carried away by the positron, the neutron and proton are no longer repelled from one another and form an entity called a 'deuteron'. A third proton can then collide with the deuteron to form a nucleus of helium-3, which contains two protons and one neutron, held together by the strong nuclear force. Finally, two helium-3 nuclei collide, forming a single nucleus of helium-4 and spitting out two protons.

Each time four protons combine in this way (or any way!) to make one helium-4 nucleus, 0.7 per cent of the original mass is released as energy. Since the Sun formed, about 4 per cent of its original stock of hydrogen has been used up in the process.

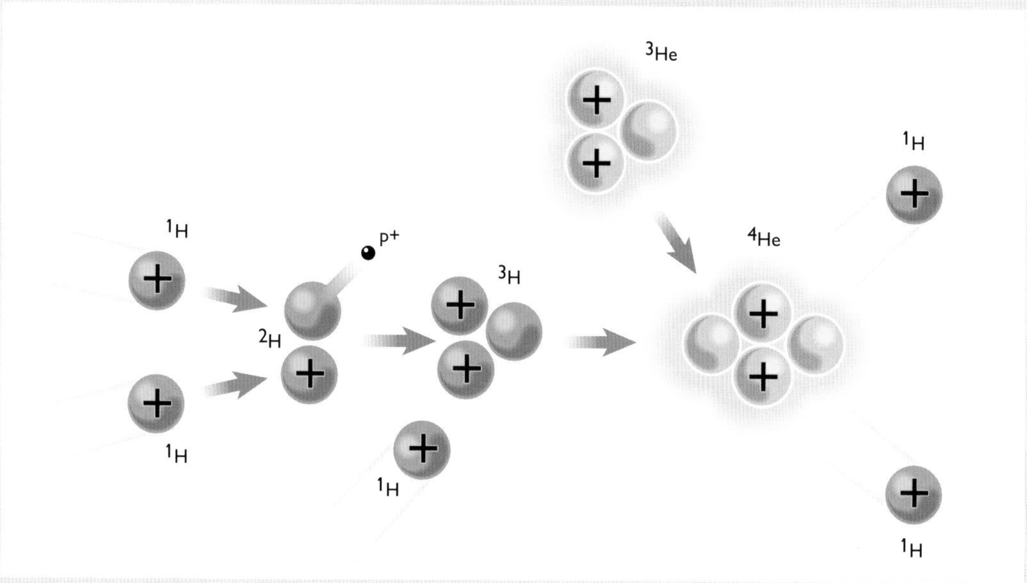

Stellar Cycles

Stars with more mass than the Sun are slightly hotter in their hearts, and generate energy through a different process, which is known as the 'carbon cycle'. (A very small amount of the Sun's energy does come from the carbon cycle, but most is produced by the p–p chain.)

The carbon cycle depends on the fact that stars today are made from the debris of previous generations of stars, and contain a smattering of heavy elements produced by nucleosynthesis. It starts when a proton enters a nucleus of carbon-12, which contains six protons and six neutrons. The resulting nucleus (nitrogen-13) is unstable and emits a positron and a neutrino as one of its protons converts into a neutron

1

1. The p–p chain, which releases heat inside the Sun.

2. A SOHO ultraviolet image of the chromosphere layer of the Sun's atmosphere, with a solar flare pictured top right.

(converting the nucleus itself into carbon-13). If a second proton penetrates this carbon-13 nucleus, it becomes nitrogen-14. A third proton will convert it into oxygen-15, which is unstable and spits out a positron and a neutron as it converts into nitrogen-15.

Now comes the crunch. If a proton penetrates a nucleus of nitrogen-15, it spits out a whole alpha particle (a helium-4 nucleus), leaving behind a nucleus of carbon-12 identical to the one that started the cycle. The net effect is that four protons have been turned into one helium-4 nucleus with energy being released along the way. As with the p–p chain, 0.7 per cent of the mass of the set of four protons is converted into energy each time an alpha particle is made.

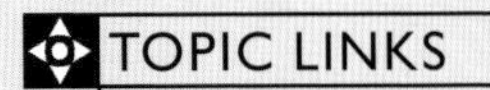

TOPIC LINKS

1.2 Star Birth
p.41 The Cosmic Squeeze

1.4 Out With a Bang
p. 67 A Bigger Blast

p.71 Black Holes and Neutron Stars

mainly of hydrogen and helium. Red giants can have masses tens of times the mass of the Sun, and diameters up to 100 times larger. Because they are so big, the gravitational pull at the surface of such a star is very weak and it is easy for material to escape into space. This is especially true when the atmosphere is being puffed in and out by variations like those we see in Cepheids.

The Fate of the Sun

The Sun itself will become a red giant in about 5 billion years' time, and at its largest it will expand to engulf Mercury and approach the orbit of Venus. You may sometimes hear or read that it will even engulf the Earth, but this prediction is not correct, because it does not take account of the fact that by that stage of its life the Sun will have lost about 25 per cent of its original mass by ejecting material into space.

The overall time that a star spends as a red giant is much less than the time it spends on the main sequence, only between 5 per cent and 20 per cent of its main sequence lifetime, depending on its mass. The Sun will be a red giant for only about a billion years, and will never get beyond the helium-burning stage. Bigger stars, though, may go through successive phases of nuclear burning, ending up with a structure like that of an onion, with different kinds of nuclear burning (and nucleosynthesis) going on in each layer.

Fading Stars

For a star like the Sun, and stars with up to a few times as much mass as the Sun, all the nuclear burning possibilities will eventually come to an end. The star will blow away its outer layers to form a planetary nebula as the core collapses and stabilizes as a solid lump of material. This dense core of star stuff starts out hot, thanks to the leftover heat of its former glory and the heat generated in its final collapse, but it is very small, about the size of the Earth. It becomes what is known as a 'white dwarf', one of the hot but faint stars that occupy the bottom left part of the HR diagram.

A white dwarf contains anything from about half to one and a half times the mass of our Sun, packed into a solid lump about the size of the Earth. One cubic centimetre of white dwarf stuff would have a mass of about 1 tonne – a million times the density of water.

If all stars faded away quietly, as the Sun is destined to do, there would be very little in the way of heavy elements in the Universe, very few planets (if any) and no life forms like ourselves. But some stars end their lives in events that play a key role in both manufacturing heavy elements and recycling them into the interstellar medium.

However, stars which start their lives with more than about eight times as much mass as our Sun have a different, and more spectacular, fate in store. They are destined to become the supernovae which trigger the next generation of star birth, allowing new stars to arise Phoenix-like from the ashes of the old.

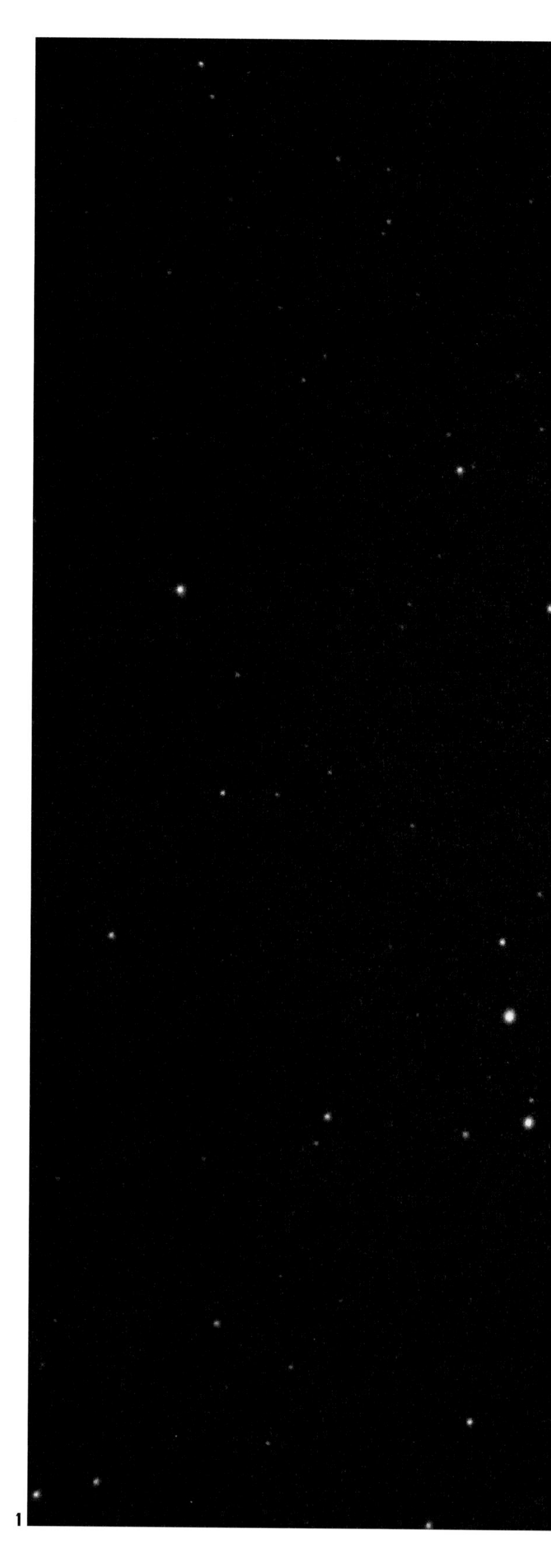

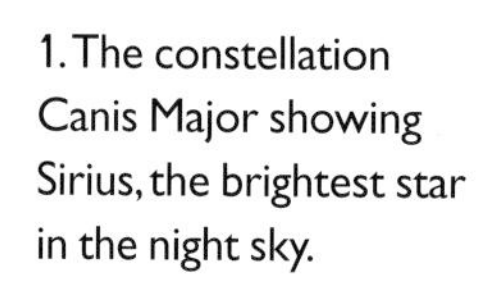

1. The constellation Canis Major showing Sirius, the brightest star in the night sky.

1.4 OUT WITH A BANG

If all stars faded away quietly, as the Sun is destined to do, there would be very little in the way of heavy elements in the Universe, very few planets (if any) and no life forms like ourselves. But some stars end their lives in explosive events that play a key role in both manufacturing heavy elements and recycling them into the interstellar medium. These supernovae have been known from observations since the 1920s, but it took a long time for their significance to be appreciated.

One of the most important discoveries of late twentieth-century science is that we are made of star dust, but it was only in the last decade of the century that astronomers worked out the details of how supernova explosions manufacture the heaviest elements and spread them through space in their death throes, providing the raw material for the formation of new planetary systems.

STARS LOCKED IN A MORTAL EMBRACE

There are two kinds of stellar explosion involved in recycling the elements, and the more common kind occurs in the most common kind of star – the ones which, unlike our Sun, are part of a binary system, with two stars in orbit around one another.

The presence of a companion star has very little effect on the early evolution of a star like the Sun, and it can happily run through its time on the main sequence in the usual way. But things get more complicated when one of the two stars leaves the main sequence and tries to become a red giant. This will happen first to the more massive of the two, since more massive stars burn their fuel faster and leave the main sequence sooner.

Swapping Matter Between Partners

If the two stars are close enough together (as is often, but not always, the case), as the more massive star swells to become a red giant it will begin to dump material from its distended atmosphere onto its companion. The gravitational pull of the companion will tug a streamer of material away from the red giant, and this extra matter will pile up on what was the less massive star in the pair, adding to its mass and speeding up its own evolution. By the time the first star has become a white dwarf, it may be lighter than its companion, which will have gained mass and become a red giant in its turn, dumping material back onto the white dwarf.

This can produce a variety of interesting phenomena, including bursts of X-rays from the point where the infalling material from the red giant creates an intense hot spot on the surface of the white dwarf. But the most important consequence of this kind of binary interaction is that it can produce repeated outbursts which blast matter off into space.

1. Artist's impression of a binary star in which a red giant is losing matter to a small companion.

1. The first x-ray astronomy satellite Uhuru, which discovered evidence for what are now thought to be black holes in our galaxy.

These explosions happen in systems where the separation of the two stars provides a more or less steady rain of gas from the red giant onto the white dwarf, instead of creating a hot spot. This material is mostly hydrogen, from the atmosphere of the giant star, and it builds up on the surface of the white dwarf at a rate of about 1 billionth of the mass of the Sun each year. But the gravitational pull at the surface of a white dwarf is so intense – tens of thousands of times greater than the gravitational pull at the surface of the Earth – that even hydrogen forms a dense layer with an intense pressure at its base. As more gas falls onto the white dwarf, this pressure increases steadily. When a thick enough layer of hydrogen has built up in this way, the pressure at the base of the layer triggers a wave of fusion activity (like the explosion of a huge hydrogen bomb), which blasts the material out into space and causes the star to flare up brightly for a short time. Then the whole process can start again as gas from the giant star continues to fall onto the white dwarf.

New Stars for Old

Such a brightening of the star is called a nova (meaning 'new'). The name derives from the fact that white dwarfs are too faint to be seen with a small telescope in their resting state, and when novas were first observed it was thought that they were literally new stars being born. During a nova, the brightness of the star may increase by 100,000 times in a few days, before fading back to its previous insignificance over a few months. The surface temperature of the star reaches about 100 million K during the outburst and it ejects the equivalent of about one ten-thousandth of the mass of the Sun in the form of heavy elements, which enrich the interstellar medium.

In a disc galaxy like our own Milky Way, about 25 novae occur each year. It is thought that all novae are produced by the binary accretion process described above and that all are subject to repeated outbursts of this kind. Some novae have actually been seen to recur in this way. The star T Corona Borealis, for example, flared up in both 1866 and 1946. All novae are believed to follow a similar pattern, but gaps between outbursts are too long for us to have seen them explode more than once since we started watching the skies.

Some white dwarfs in binary systems suffer a more extreme fate. A white dwarf is made of atomic nuclei (atoms stripped of their electrons) jostling together, in a sea of electrons, with one electron for every proton present. This is very nearly the most compact form in which matter exists. But the rules of quantum physics tell us that there is a more dense form of matter, which could be achieved if each proton absorbed one electron and became a neutron. If that happened, all the matter would shrink down to become a single ball of neutrons, like a single huge atomic nucleus.

Matter at its Most Extreme

The pressure required to make this happen is immense, but it will be reached if the mass of the white dwarf exceeds 1.4 times the mass of the Sun. So if a white dwarf with just a bit less than this critical mass is quietly accreting matter from a companion star, when its mass reaches this limit the inside of the star will collapse as the material in its core is turned into a ball of neutrons. As the star collapses, a huge amount of gravitational energy is released in the form of heat, triggering a wave of nuclear reactions in the material the star is made of. The matter the star contains is blasted away into space. Such an event is called a supernova, because it far exceeds the outburst of a nova. In a supernova, a single star can briefly shine as brightly as a whole galaxy of billions of ordinary stars.

The kind of supernova produced by the disruption of a white dwarf in a binary system is called a 'Type I supernova'; because all Type I supernovae are produced in the same way, from white dwarfs with exactly the same mass, they all have the same brightness. This makes them good standard candles to use in measuring distances to other galaxies. The other important thing about Type I supernovae is that they spread huge amounts of heavy elements through space. A Type I supernova laces the interstellar medium with about half to one solar mass of iron, plus 12–15 per cent of a solar mass of oxygen, and a smaller amount of other heavy elements. But even this is not the most spectacular way in which a star can die.

THE STAR OF BETHLEHEM

One thing we know for sure about the birth of Jesus Christ is that it did not occur in the year 1. It was only in the sixth century (by the modern calendar) that the idea of counting time from the birth of Jesus was suggested by the Roman scholar Dionysus Exiguus, and he simply made a mistake in counting back to the all-important year. We know this, because King Herod, who played such an important part in the Nativity story, died in 4 BC. But while Herod is an historical fact, is there any truth in the Biblical story of the Star of Bethlehem?

It so happens that Chinese astronomers were keeping records of unusual events in the heavens at that time, and these records include a report of the presence of a 'guest star' in the constellation Capricorn in March of the year we now call 5 BC.

The date fits neatly with the Biblical account – not only the year (before Herod died) but the season (March would be a likely time for shepherds to be out tending their flocks in the lambing season). Intriguingly, a faint nova was seen in the same part of the sky in 1925, and since all novae are thought to be recurrent, it may be that this was the star observed by the Magi. Whether it was or not, the expanding shell of debris from the nova observed by the Chinese should soon be detectable to the latest telescopes, and that will pin down its exact location. Even without that confirmation, however, some astronomers are convinced that the evidence is strong enough to pin down the time of the birth of Jesus not 25 December in the year 1, but late March in the year 5 BC. Which, among other things, means that everybody celebrated the end of the second millennium after His birth five years late!

1

2

1 and 2. Supernova 1987 A (left) compared with the same star before it exploded (right).

A BIGGER BLAST

Type II supernovae occur in massive, young stars rich in heavy elements produced by nucleosynthesis. These stellar explosions mainly occur in the spiral arms of disc galaxies (▷ p. 22), because the stars involved are so massive they do not have time to move far from their place of birth before they die. They also occur in other regions where starbirth has been triggered. This happens, for example, when clouds of gas and dust in a relatively quiet galaxy are disturbed by the tidal forces created when another galaxy passes nearby, and collapse to form new stars. Type II supernovae release even more energy when they explode than Type I supernovae, but they are not as bright to conventional telescopes, because most of this energy is released in the form of invisible particles called neutrinos. A Type II supernova produces in a few minutes about 100 times as much energy as the Sun will radiate over its entire lifetime of 10 billion years or more.

The more massive a star is, the more quickly it burns its fuel, and the shorter its life. The progenitors of Type II supernovae can have masses tens of times that of our Sun, but to give an example of how they are formed we must look at the evolution of a star which starts out with a little less than 20 solar masses of material. It is thought to be the mass of a supernova that was seen to explode in the Large Magellanic Cloud in 1987, the supernova known as SN 1987A.

Running Out of Fuel

Such a star has to burn nuclear fuel so fiercely to hold itself up that it shines 40,000 times brighter than our Sun and spends only 10 million years on the main sequence. Burning helium as a red giant supplies it with energy for only another million years or so, and then it runs through the remaining possibilities offered by nuclear fusion at an ever faster rate. Converting carbon into a mixture of oxygen, neon and magnesium provides energy for 12,000 years; burning neon and oxygen sustains it for another 16 years; and fusion of silicon nuclei to make iron-family nuclei keeps it going for about a week. During that last week of its life as a more or less stable star, the inner core of the giant star is like a series of Russian dolls, with each of these nuclear

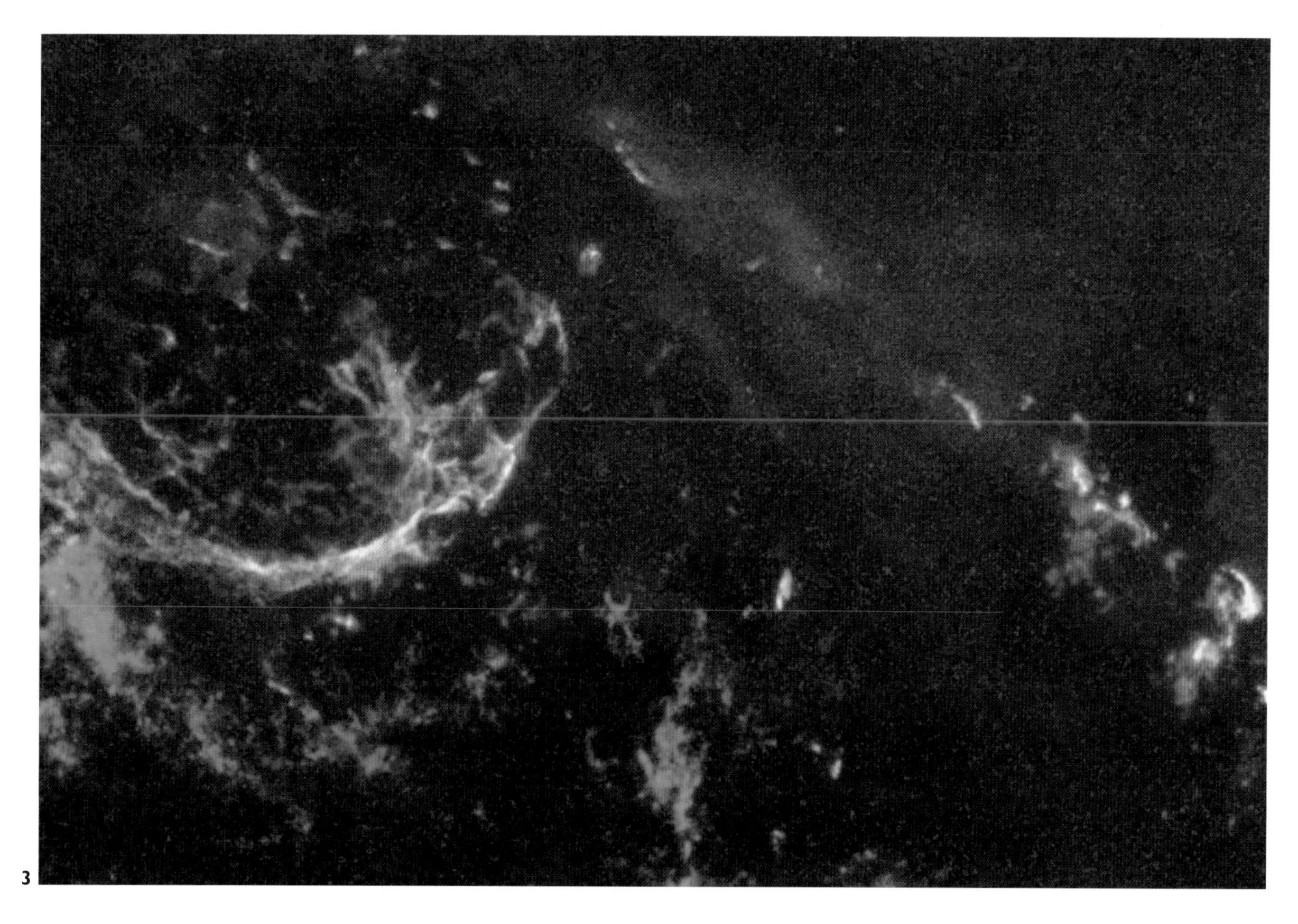

3. The remnant of a supernova that exploded long ago in the Large Magellanic Cloud. The colour indicates the presence of large amounts of oxygen.

1. One of our nearby irregular galaxies, NGC 1313.

2. Relative sizes of stars. To help put this into perspective, a white dwarf is about the same size as Earth.

fusion processes going on one inside the other.

Once the silicon in the core has been converted into iron-family elements, there is no longer a source of energy to provide pressure to hold the star up against its own weight. With the rug pulled from under it, the star collapses in spectacular fashion, converting gravitational energy into a heat so intense that it breaks the heavy nuclei apart. This produces a pressure so great that it forces electrons to combine with protons to make neutrons. The inner core of the star collapses in a few seconds from a sphere of star-stuff bigger than the Sun into a ball of neutrons about 20 km across. This leaves the outer layers of the star, with perhaps 15 times as much mass as our Sun (remember that it will have lost a lot of its original mass during its time as a red giant), plunging inward at about a quarter of the speed of light. But the formation of the neutron star produces a shock wave which ripples out from the core in a kind of rebound, rapidly followed (really rapidly, at almost the speed of light) by a blast of neutrinos. One neutrino is released for every proton that combines with an electron to make a neutron.

The combination of the shock wave and the neutrino blast turns the collapsing outer layers of the star (all 15 solar masses) inside out, and sends them hurtling out into space to form a rapidly expanding, glowing cloud of gas – a supernova remnant.

A Blast of Oxygen

Whereas Type I supernovae eject huge amounts of iron into the interstellar medium, most of the iron in a Type II supernova is converted into neutrons in the core collapse. The material that gets thrown out by a Type II supernova is very rich in oxygen – perhaps as much as 1.5 solar masses of oxygen for a 20 solar mass supernova – but it also contains all the heavy elements produced by stellar nucleosynthesis, plus a smattering of elements even heavier than iron, produced in the extreme conditions of the supernova itself (particularly in the shock wave). This is where things like gold, zinc, uranium, and everything heavier than iron originate.

The amount of very heavy elements produced, however, is tiny compared with the amount of hydrogen and helium in the Universe. Remember that everything else (all the nuclear matter that is not hydrogen or helium), put together amounts to less than 1 per cent of the matter that is. Of that everything else, all of the nuclei of everything heavier than iron put together add up to less than one thousandth as much matter as all the nuclei from lithium (the third lightest element) to iron family elements put together. Without Type II supernovae, the very heavy elements would not exist at all. Life might still exist, but there would be no radioactivity and no fission bombs.

A thimbleful of neutron star stuff would contain as much mass as there is in the bodies of all the people on Earth put together.

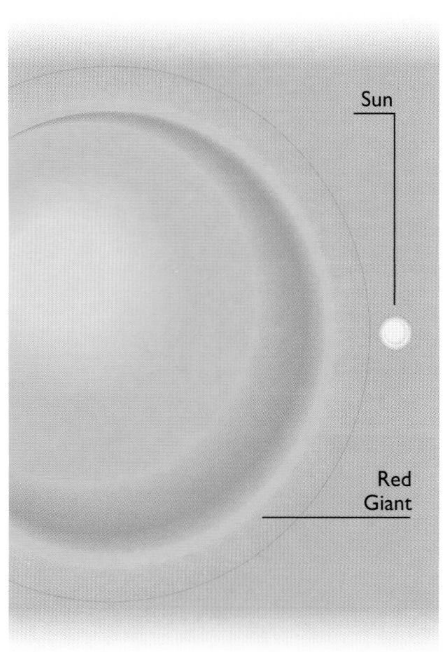

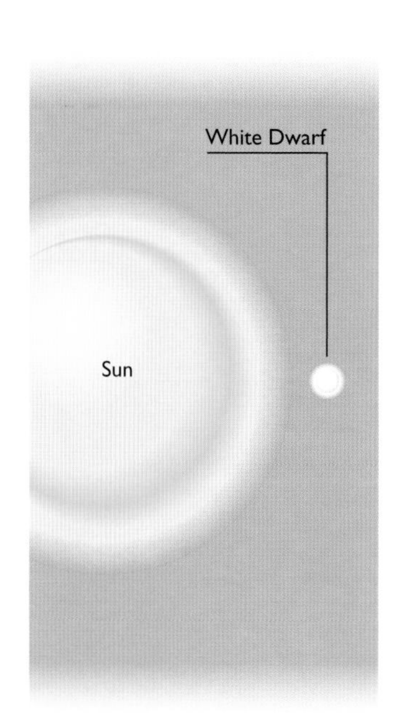

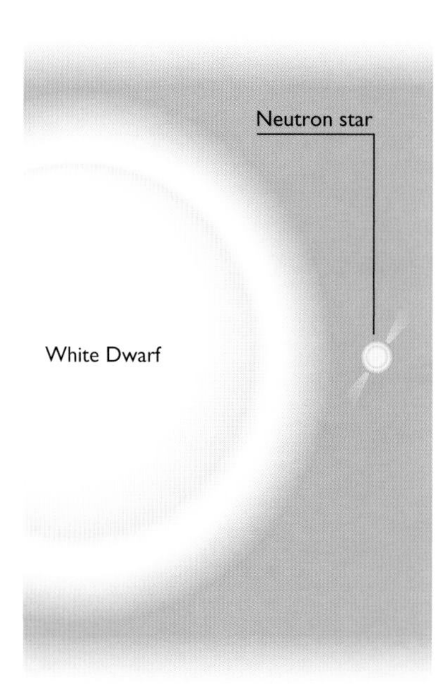

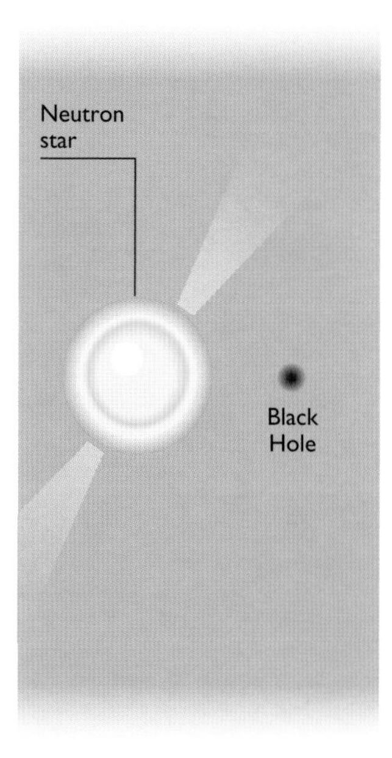

2

1. A detailed view of the Crab Nebula shows how the nebula is still being affected by the activity of the central neutron star.

2. The magnetic field of a spinning neutron star forces radiation out in two beams, like a lighthouse. Such an active, spinning neutron star is called a pulsar.

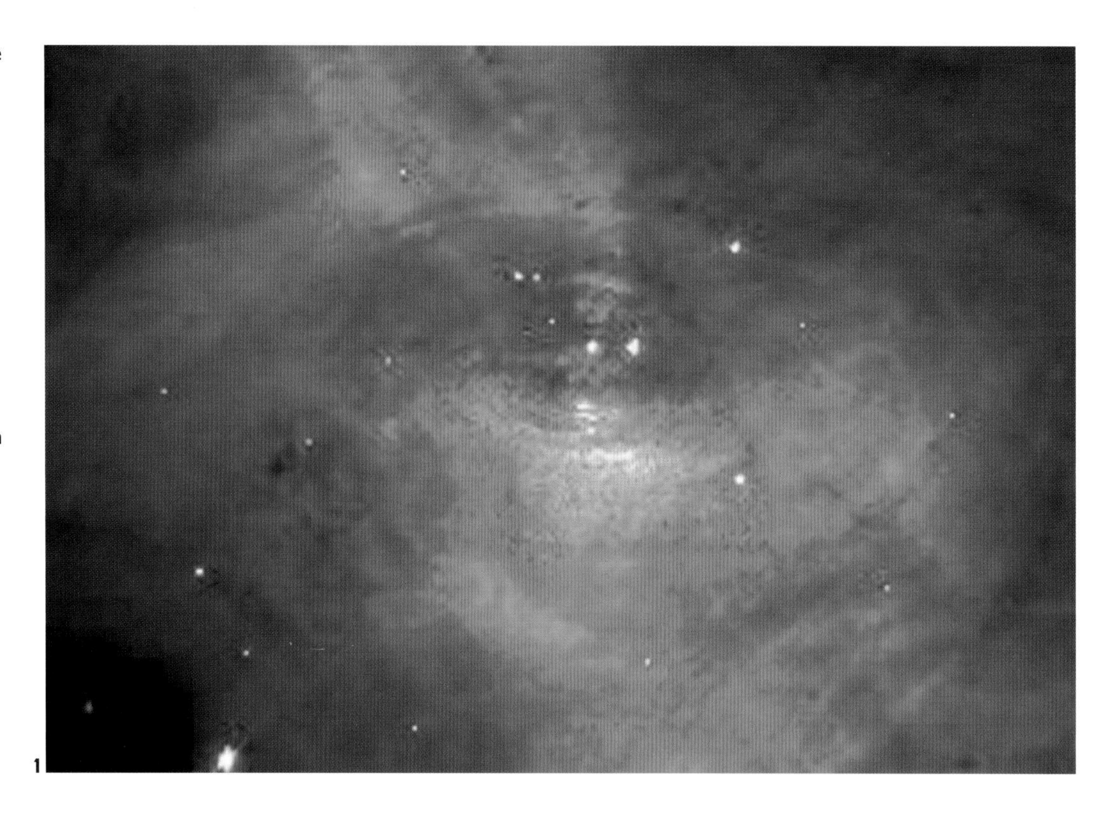

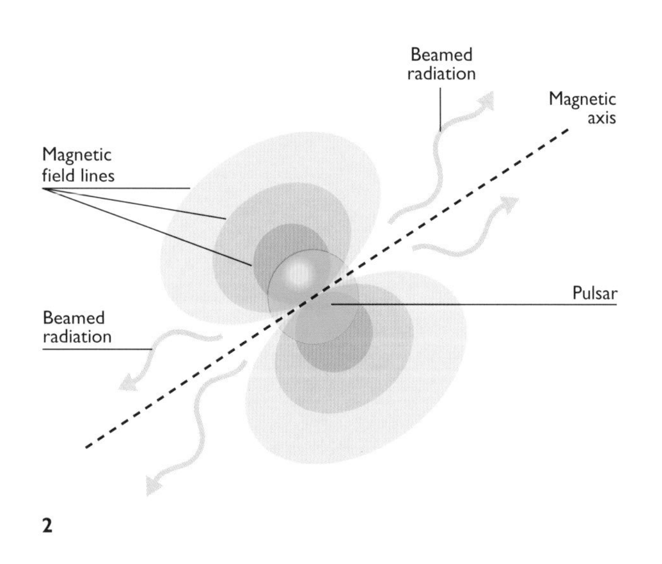

PULSARS

A spinning, magnetic neutron star produces beams of radio waves which sweep around as the star rotates. When these beams happen to flick past the Earth (like the beams of a cosmic lighthouse) they produce regular, rapid blips of noise in radio telescopes. Such objects are called pulsars. The first known pulsars were discovered by accident in 1967, by a team at the University of Cambridge that had built a new kind of radio telescope to look at flickering in the radio output of quasars. When theorists realized that pulsars must be spinning neutron stars, it opened the way for a new wave of investigation of very dense objects (neutron stars and black holes) in a classic example of how observation and theory combine in the exploration of the Universe.

Roughly a thousand pulsars have now been discovered, and the number is growing. The magnetic field of a pulsar is about 1000 million times as strong as the magnetic field of the Earth. Most of them spin once every second or so; the slowest pulsar has a period of about 4 seconds, but the fastest yet discovered spins on its axis more than 600 times a second. Imagine a ball of stuff the size of Mount Everest but containing as much mass as our Sun, spinning once on its axis every 1.6 milliseconds and you have some idea what a pulsar is like.

BLACK HOLES AND NEUTRON STARS

The remnant left behind by such a supernova is almost as interesting as the supernova itself. It is a star made almost entirely of neutrons – a neutron star – matter packed into the most compact state possible, with a density the same as that of an atomic nucleus. If such a neutron star had the same mass as our Sun, it would have a diameter of only about 10 or 20 kilometres, the size of a large mountain on Earth. This is far smaller even than a white dwarf which packs about a solar mass of material into a sphere about the same size as the Earth. The density of matter in a neutron star is about 1 million times the density of a white dwarf. If it could be magically transported to Earth and kept in its superdense state, a cubic centimetre of a neutron star would weigh about 100 million tonnes.

Under the extreme conditions of pressure that occur at the heart of a Type II supernova, neutron stars with as little as one-tenth of the mass of our Sun can form. But any neutron star with less mass than this that was formed in such a blast would expand as the pressure was released and turn into an unusual low-mass white dwarf (with some of the neutrons converting themselves into protons).

In the Realm of Speculation

A few astronomers speculated about the existence of neutron stars in the 1930s, soon after the neutron itself was discovered. One team, Walter Baade and Fritz Zwicky, even speculated (in 1934) that the only way to explain the energy output of a supernova would be by the transition of an ordinary star into a neutron star, releasing huge amounts of gravitational energy. But whereas white dwarfs had been identified in 1934, nobody had ever seen a neutron star, and most astronomers could not accept what the equations of physics were telling them. They didn't believe that such superdense objects really existed. Baade and Zwicky's speculation was not taken seriously by most of their colleagues for three decades, until the accidental discovery of pulsars, which were explained as rapidly spinning, magnetic neutron stars.

The Earth would have to be squeezed to the size of a large pea in order to become a black hole.

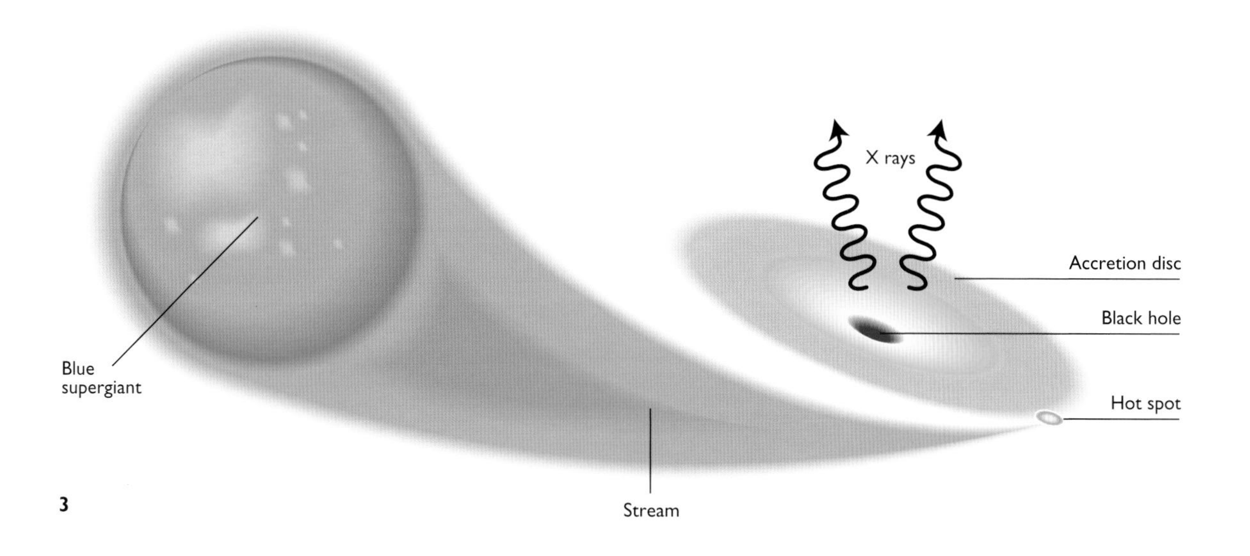

3. When matter from a giant star streams onto a black hole companion, the material gets hot and emits x-rays.

THE SEARCH FOR BLACK HOLES

Even after the discovery of pulsars started astronomers thinking about the possibility of black holes, there was no proof that they existed. But at the beginning of the 1970s, a satellite called Uhuru pinpointed the position on the sky of an X-ray star (▷ p. 64) called Cygnus X-1. The positioning was accurate enough for optical astronomers to identify the star with their telescopes. (Cygnus X-1 gets its name because it is the brightest X-ray star in the direction of the constellation Cygnus, the swan.)

A Dying Swan

When astronomers studied the source of the X-rays from Cygnus, they found that the X-rays were coming from a point near a blue star called HDE 226868, but not from the star itself. The star and the X-ray source orbit around each other once every 5.6 days and the orbit corresponds to an object with a mass of about 20 times the mass of the Sun. This means that the X-ray source cannot be either a white dwarf or a neutron star, and if it were an ordinary star that big, held up by nuclear fusion, it would be bright enough to see (nearly as bright as its blue companion). Theory and observations combined to show that it could only be a black hole – a dead (or dying) star. Several similar objects are now known, and they are referred to as 'stellar mass black holes', because they have masses similar to those of stars.

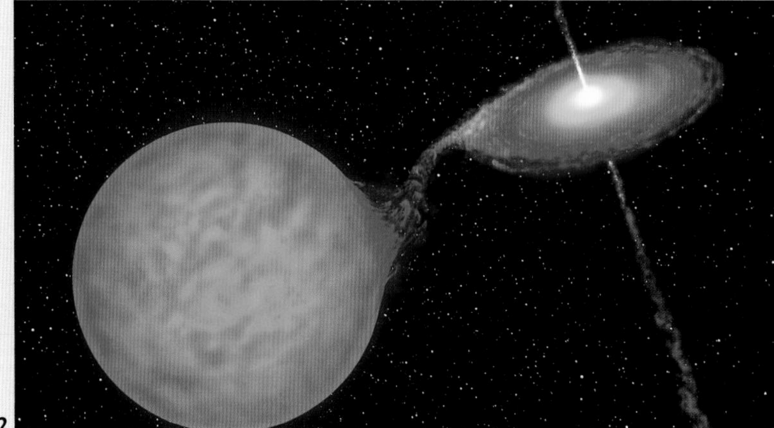

2

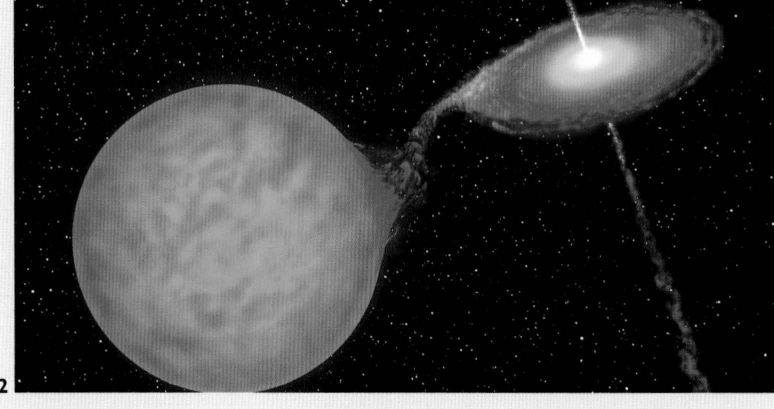

3

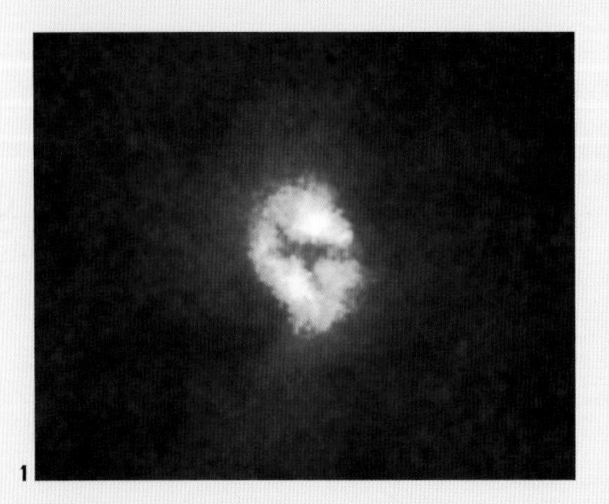

1

Stellar mass black holes are difficult to find, because they only show up if they are in orbit around another star, stripping matter away from it and swallowing it in a messy fashion. An isolated black hole is, indeed, black. But using our understanding of how stars evolve, Roger Blandford, of Caltech, has estimated that there may be 100 million isolated black holes scattered across our Milky Way galaxy, and that the nearest one may be only 5 parsecs (15 light years) away.

Bigger Black Holes

From the late 1960s onward, astronomers speculated that quasars must be what are known as 'supermassive black holes', with masses roughly 100 million times the mass of the Sun, because the only way that enough energy can be liberated to explain the output of a quasar is if it is a large black hole swallowing matter from its surrounding galaxy. This was finally proved in the 1990s, when the Hubble Space Telescope obtained images which show the discs of material swirling around some of these black holes. The size of these discs and the speed with which they move (revealed by the Doppler

4

1. Actual photograph of the centre of a system like the ones illustrated here. This is a black hole at the heart of the galaxy M51 taken by the Hubble Space Telescope.

2. Illustration of a super-massive black hole ejecting jets of material from its poles.

3 and 4. Close-up views (artist's impression) of black hole activity.

effect – ▷ p.120) tells us how big the central black holes in these galaxies are.

In a typical example, the galaxy NGC 7052, there is a disc of material 1100 parsecs (3700 light years) in diameter swirling round a black hole with a mass 300 million times the mass of our Sun. The disc itself contains 3 million solar masses of material – enough to keep the quasar shining for 3 million years, because it swallows only one solar mass a year.

Bending Space

The general theory of relativity explains gravity as a result of space being bent by the presence of matter. In a neat aphorism, physicists say that matter tells space how to bend; space tells matter how to move. In effect, objects moving under the influence of gravity roll along the valleys in a hilly landscape of curved space (strictly, curved spacetime). Using this image, flat spacetime, without any matter in it, is like a stretched flat sheet of rubber (or a trampoline). A heavy weight placed on the stretched sheet makes a dent, and anything rolling across the sheet follows a curved path around the dent. If the object is heavy enough, it stretches the fabric so much that it makes a deep pit with vertical sides, from which nothing can escape – a black hole.

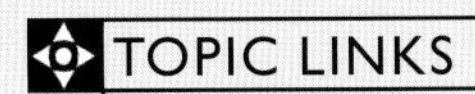

1.4 Out with a Bang
p. 71 Black Holes and Neutron Stars
p. 75 Proof of the Pudding

2.1 The Big Bang
p. 87 The Singular Beginning

4.3 Into the Unknown
p. 217 Beyond the Black Hole
p. 230 The Evolution of Universes

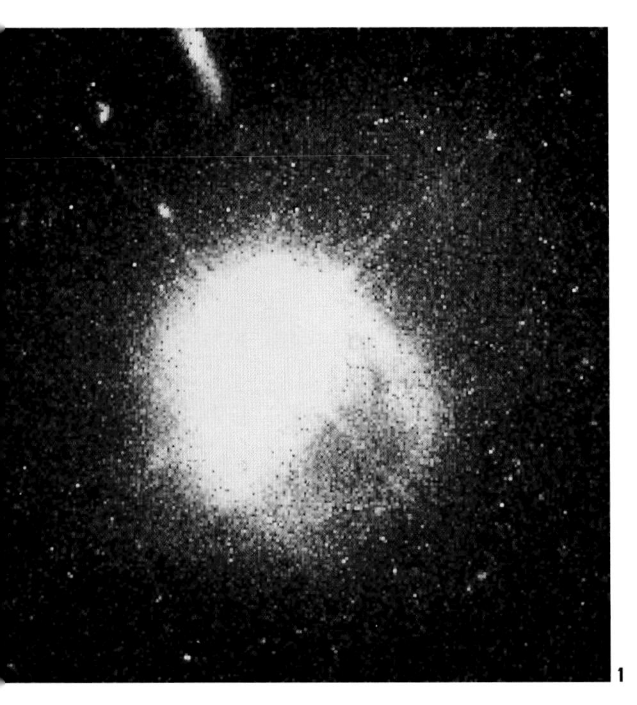

1

Singularities

The discovery of pulsars meant that the equations which said neutron stars exist had to be taken seriously. That meant that another, even stranger, prediction of those equations also had to be taken seriously. At the end of the 1930s, Robert Oppenheimer and George Volkoff had shown that the equations which say neutron stars exist also say that there is an upper limit to the amount of mass a neutron star can have. The exact value of this upper limit depends on subtleties in the equations, which the experts still argue about, but it is about three times the mass of the Sun. What would happen to any neutron star that tried to form with more mass than this, or which started out with less mass but crossed the line as a result of accreting material from a companion? The equations said that it would shrink down indefinitely, towards a single point – a singularity.

But on the way to a singularity, a collapsing object would disappear from view, because the gravitational pull at its surface would become so intense that nothing, not even light, could escape. The work of Robert Oppenheimer and his colleagues was not followed up at the time because of the Second World War. Oppenheimer himself worked on the Manhattan Project – to build the first nuclear bomb.

In fact, collapsed objects like this had already been described, using the equations of Albert Einstein's general theory of relativity, by Karl Schwarzschild, in the second decade of the twentieth century. In terms of relativity theory, the collapsed object is pinched off from our Universe because space itself (strictly speaking, 'spacetime') has been bent (or warped) around it, making a hole in space. The interior of the hole is, in effect, a separate, self-contained universe. These objects were only given their now-familiar name, 'black holes', in 1967, just after the discovery of pulsars made everyone begin to take the idea seriously.

Getting to Grips with Black Holes

Any object will form a black hole if it is squeezed into a small enough volume. For a particular mass, the crucial radius at which this occurs is called the Schwarzschild radius, and this becomes, in effect, the radius of the black hole. Once anything is squeezed within its Schwarzschild radius, it will collapse to a singularity, leaving a black hole with that radius as its imprint on our Universe, like the grin on the face of the fading Cheshire cat. For the Sun, the Schwarzschild radius is just 2.9 km, which shows how close neutron stars (with radii of about 10 km) are to becoming black holes. For the Earth, it is only 0.88 cm. But black holes need not necessarily be associated with very high densities. An object with the same density as the Sun (or water) would be a black hole if it were about as big across as our Solar System.

There is now direct proof that black holes exist. Even though nothing can escape from a black hole (so it cannot be seen directly), there may be a great deal of activity going on just outside the Schwarzschild radius (which is also known as the 'event horizon'). If a black hole with a radius of a few kilometres is in orbit around another star, stripping matter away from it and swallowing it up, there will be an intensely swirling maelstrom of stuff orbiting around the black hole in a disc and being sucked in. Gravitational energy released as the matter falls into the black hole will heat this disc to the point where it radiates X-rays. Such X-ray stars have been detected orbiting ordinary stars, and their masses, sometimes exceeding 10 solar masses, have been measured from studies of their orbits.

Gamma ray bursts from space were first identified in the late 1960s by American satellites designed to look for gamma rays from any secret nuclear bomb tests being carried out in Russia.

There is no doubt that some of these X-ray stars are black holes.

Much more massive black holes (perhaps containing 100 million solar masses of material, and as big across as our Solar System) are also thought to reside at the hearts of galaxies, where the energy released as they swallow matter makes the gas around some of them shine more brightly than the rest of the galaxy put together. They are known as quasars. To give you some idea of the enormous amount of energy that can be released when matter falls into a black hole, the energy output of a quasar can be maintained by swallowing only about one solar mass of material each year.

Proof of the Pudding

The best evidence that black holes exist came in the 1990s, when astronomers identified the sources of intense bursts of gamma rays detected by instruments on board satellites orbiting the Earth. These gamma ray bursts had long been known, but it was only at the end of 1997 that astronomers were first able to identify the source of one of these bursts using ordinary telescopes. It turned out to be in a galaxy more than 3 billion parsecs (10 billion light years) from Earth. In order to produce a burst of gamma rays visible to our detectors that far away, for a few seconds the object had radiated as much energy as every star in every galaxy in the visible Universe put together. In a region about 150 km across, it briefly produced conditions like those which existed in the Big Bang itself. The only way in which to generate such enormous outbursts of energy in such a short time is in a kind of super supernova (a hypernova?) where the core collapse does not stop at the neutron star stage, but goes all the way to a black hole. The power of a quasar (as bright as several hundred billion Suns) comes from swallowing about one solar mass of material in a year; the power of a gamma ray burster comes from swallowing several solar masses of material in a few seconds. It is the ultimate form of stellar death.

2

1. The quasar PKS2349 photographed by the Hubble Space Telescope.

2. Radiation coming from a region of space where a black hole has collapsed is like the fading grin of the Cheshire cat.

2

THE FATE OF THE UNIVERSE

2.1 THE BIG BANG

One of the greatest achievements of the human intellect was the discovery in the twentieth century that the Universe as we know it originated in a hot, superdense state at a definite moment in time (a beginning, if you like), that it has been expanding ever since, and that we can work out when time began – about 14 billion years ago. It is pleasing too that the age of the Universe, determined from cosmology, closely matches the ages of the oldest stars determined by astrophysics.

There is overwhelming evidence in support of both the idea of the Big Bang, as it is known, and when it occurred. At the beginning of the twenty-first century, instead of resting on their laurels, cosmologists are now attempting to answer the other big questions about the Universe: where it is going and how it will all end. The first tantalizing clues that we might soon be able to give definite answers to these questions came at the end of the twentieth century.

Previous page. The Hubble Deep Field. Galaxies seen by light that left them more than 10 billion years ago.

THE LAW ACCORDING TO HUBBLE

The most important thing we know about the Universe is that it is expanding, that galaxies (or rather whole clusters of galaxies) are getting farther apart as time passes. We cannot see the distance increasing between them, because the distance scales and time scales involved are so huge. Even if we watched for a million years, we would scarcely be able to detect the expansion of the Universe directly. But we know for sure that the Universe is expanding, because we can measure both the distances to many galaxies and the speeds with which those same galaxies seem to be receding from us.

The key discovery is that there is a very simple relationship between these two properties, telling us that the apparent velocity recession is proportional to the distance of a galaxy from us. This is known as Hubble's Law, and the constant of proportionality in Hubble's Law is called the Hubble Parameter (or sometimes, Hubble's Constant); it is denoted by the letter H. Hubble's Law does not mean, however, that we are at the centre of the Universe. It is the only velocity/distance law (except for a situation in which nothing is moving at all) which holds wherever you are in the Universe.

Everything is receding from everything else, like raisins being carried away from one another in the rising dough of a raisin loaf as it

To a cosmologist, a galaxy like the Milky Way, containing hundreds of billions of stars, is the *smallest* thing in the Universe worth taking account of.

1. The Andromeda galaxy, also known as M31, is the nearest large spiral galaxy to our own.

cooks. Whichever galaxy you sit on, it will look as if all the other galaxies are rushing away from you with speeds proportional to their distances.

Beyond the Local Group

The first measurements of distances to galaxies beyond the Milky Way were made at the end of the 1920s, using the Cepheid distance scale (▷pp. 28–9). But even with the aid of the best telescopes that were available before the 1990s, it was not possible to detect Cepheids in more than a handful of the very nearest galaxies to our own. However, this was enough to show that the Milky Way and the Andromeda galaxy (also known as M31) are the two largest members of a small group of galaxies which also includes the Large and Small Magellanic Clouds, and is collectively known as the Local Group. The Local Group is actually a very small cluster of galaxies – other clusters of galaxies contain hundreds or even thousands of individual galaxies. In a cluster, the individual members are held together by gravity, and move around within the cluster like individual bees moving within a swarm. But the cluster as a whole takes part in the expansion of the Universe (the swarm of bees moves as a unit).

Right up until the 1990s, measuring distances to galaxies beyond the Local Group depended on using so-called 'secondary indicators' calibrated within the Local Group. Because the distances to galaxies in the Local Group (in particular, the distance to the Andromeda galaxy) are known from the Cepheid method astronomers can study the apparent brightnesses of bright objects in these nearby galaxies and use the known distances to work out how bright they really are. This makes it possible to calibrate the brightness of things like globular clusters, supernovae, and the huge star-forming clouds known as 'HII regions'. Then, by identifying

THE OLDEST THINGS IN THE UNIVERSE

Until the middle of the 1990s, astronomers were slightly embarrassed to admit that their best estimates of the age of the Universe came out at slightly less than their best estimates of the ages of the oldest stars. Obviously, the Universe must be older than the stars it contains, but measurements of the Hubble Parameter using ground-based telescopes gave rather rough and ready estimates. They showed that the Universe was about 10–12 billion years old, while the oldest stars were thought to be 14–15 billion years old. The astronomers weren't too perturbed, however, because both measurements were difficult and plagued with uncertainties, and they expected one or both of them to need adjustment when telescopes were flown above the obscuring layer of the Earth's atmosphere.

In fact, both of them needed adjusting, and in the 'right' way. In the second half of the 1990s, data from the Hubble Space Telescope (left) showed that the Hubble Parameter is a little smaller than had been thought. This meant that the best estimate for the age of the Universe became 14 billion years. At almost the same time, data from the HIPPARCOS satellite showed that some of the stars used to calibrate stellar ages were a little further away than had been thought, so they must be a little brighter than was thought in order to look as bright as they do to us. And that implies they must burn their fuel faster than anyone had thought, so they haven't taken as long to get to their present state. The best estimate for the ages of the oldest stars came down, as a result, to 12–13 billion years.

1. The way open clusters of stars move through space is like a swarm of bees moving through the air.

2. The globular cluster 47 Tucanae.

the same sort of objects in galaxies beyond the Local Group, and measuring the brightness (or faintness) of those objects compared with their counterparts in the Andromeda galaxy, it is possible to estimate the distances to more remote galaxies.

The key step in all this is working out the distance to the Virgo Cluster, which contains about 2500 galaxies spread over a spherical volume of space the centre of which is about 17 million parsecs (17 Megaparsecs, or roughly 55 million light years) away from us. The Virgo Cluster contains so many different galaxies that it provides ample secondary indicators that can be used (after they have been calibrated from the measured distance to the cluster) to work out distances to even more remote galaxies.

But the distance to the Virgo Cluster itself was only measured really accurately in the 1990s, when the Hubble Space Telescope was able, for the first time, to pick out individual Cepheids in some of the galaxies in the cluster. That is why the distances to galaxies beyond the Local Group were only finally pinned down in the last decade of the twentieth century. And that meant they could be reliably compared with the velocity measurements to work out the Hubble Parameter, and therefore the age of the Universe, which depends on the Hubble Parameter.

Galaxies on the Move

The 'velocities' of galaxies associated with the expansion of the Universe are measured in terms of redshift , and in some ways they behave like Doppler shifts. But they are not Doppler shifts, and they are not really velocities (this caused a lot of confusion at one time, but is quite easy to understand now that cosmologists have a clear idea of what is going on).

Redshifts in the light from a few other galaxies (then called 'nebulae') were first identified in the second decade of the twentieth century. But they were first studied systematically and put into their modern cosmological context by Edwin Hubble and

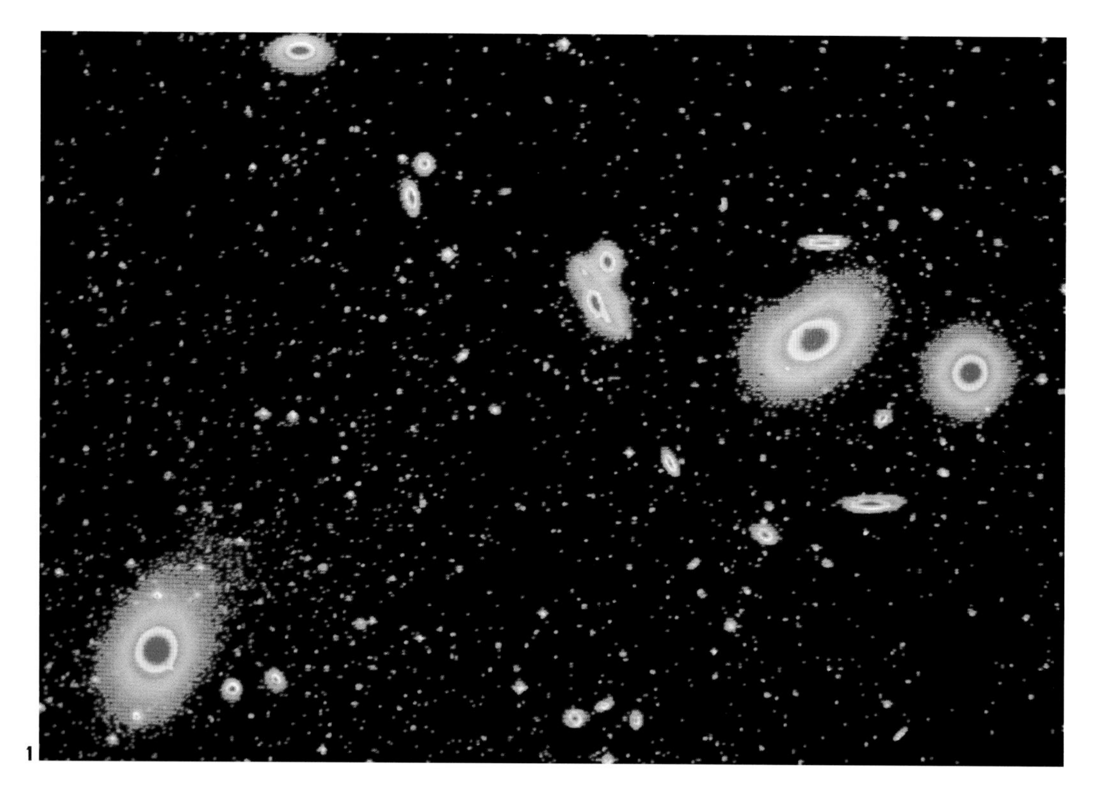
1

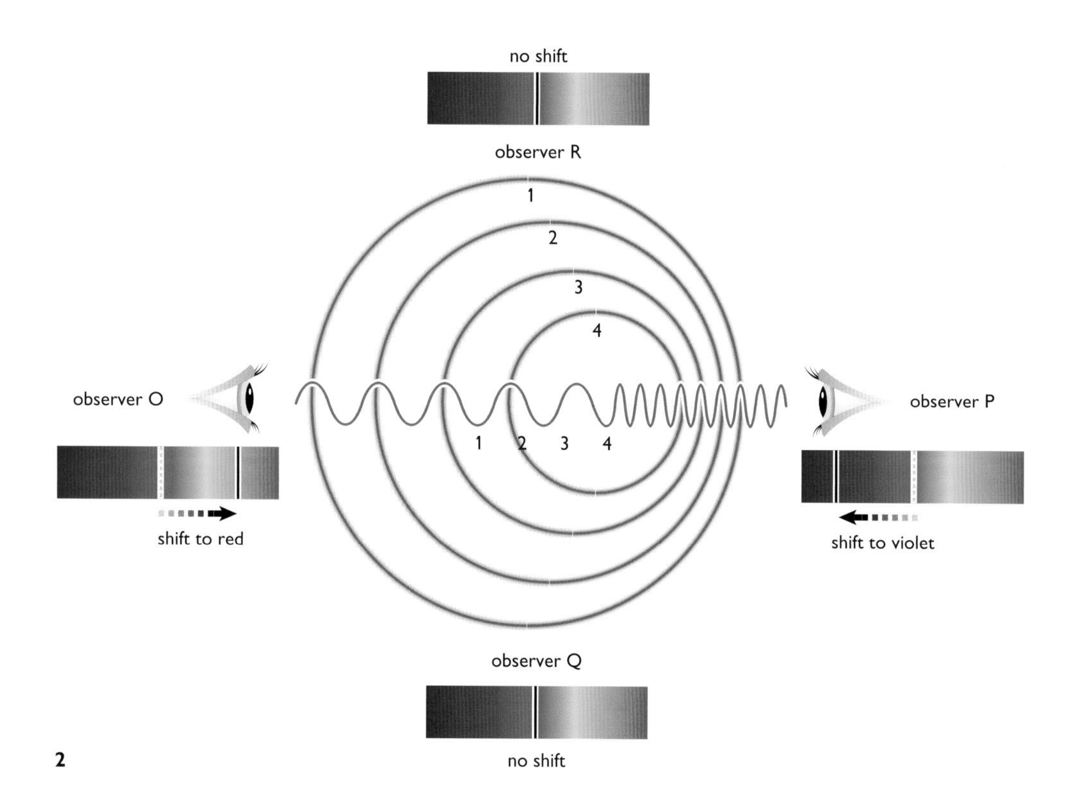

2

Milton Humason, working at the Mount Wilson Observatory in California, in the 1920s and 1930s. They found evidence that the redshift of a galaxy (provided it is outside the Local Group) is proportional to its distance from us.

Because a Doppler redshift would mean that an object is moving away from us through space, at first it was thought that this cosmological redshift was because the galaxies were hurtling apart through space, like pieces of shrapnel from the explosion of a huge bomb. But the recession of the galaxies from one another was quickly explained in terms of Albert Einstein's general theory of relativity, which describes the behaviour of space and time under the influence of matter. Einstein's theory says that space itself (strictly speaking, spacetime) is flexible, and can be stretched and squeezed.

When the equations of Einstein's theory are used to provide a description of the whole Universe (what cosmologists and mathematicians call a 'model' of the Universe), they say that it is possible for either stretching or squeezing to happen naturally, but that it is *not* possible for the Universe to stand still. Since we see other galaxies moving away from us, we know that what is happening is a stretching of space. The space between the galaxies (strictly, between clusters of galaxies) is expanding and taking galaxies along for the ride. The galaxies are not moving through space, so this is not a velocity. But because space is stretching while light from distant galaxies is travelling through it, the light gets stretched to longer wavelengths – it is redshifted. And this redshift mimics the Doppler Effect.

The distances between clusters of galaxies are getting bigger today, at a rate we can measure. This means that clusters of galaxies used to be closer together. If we could wind the expansion backwards for long enough, all

3

1. This false-colour image of part of the Virgo cluster of galaxies, highlights the nuclei of galaxies in red and their outer regions in blue.

2. The Doppler Effect squashes up light from an object moving towards you, stretches light from an object moving away, but it has no effect when the object is viewed from the side.

3. The Hale telescope, Mount Palomar, California.

the galaxies would be touching one another. Go back further and they would be squashed together into one hot lump. This is the origin of the idea that the Universe was born in a hot Big Bang at a definite moment in time. We can calculate when this happened – about 14 billion years ago – from Hubble's Law. But is there any other evidence that the Big Bang really happened?

Microwaves from the Birth of Time

The discovery that convinced many people that the Big Bang really happened came in 1965, when two researchers at the Bell Laboratories in the United States, Arno Penzias and Robert Wilson, discovered a faint hiss of radio noise coming from all directions in space. This is exactly like the kind of static you hear on a badly tuned radio receiver – indeed, some of that static is radio noise from the depths of space.

This 'cosmic microwave background radiation' had been predicted, using Big Bang theory, by George Gamow and his colleagues in the late 1940s. But Penzias and Wilson didn't know that when they made their discovery, by accident, while testing a new radio telescope.

The Big Bang theory says that the furthest back we can ever 'see' using any kind of electromagnetic radiation is the time when the whole Universe was as hot as the surface of the Sun is today. Before that time it was so hot that electrons were stripped from their atoms, leaving a mixture of negatively charged electrons and positively charged nuclei (called a 'plasma'). Because electromagnetic radiation interacts with charged particles, under those conditions the electromagnetic waves are bounced around like crazy and thoroughly mixed up. This is why we cannot see inside the Sun. For the same reason, we cannot see further back

★ The name 'Big Bang' to describe the birth of the Universe was coined by cosmologist Fred Hoyle (now Sir Fred) in a BBC radio broadcast in the 1940s.

1. A representation of the COBE satellite in orbit.

2. George Gamow, one of the founders of the Big Bang theory.

1

towards the Big Bang than the time when the entire Universe was as hot as the surface of the Sun is now.

The Very First Light

Once the whole Universe cooled to the temperature of the surface of the Sun today, electrons and nuclei combined to make neutral atoms. By and large, these do not interact with light – they will, but only if the wavelength of the light is just right to match the steps in their spectroscopic energy levels (▷ pp. 18–19). This meant that the light from the Big Bang could at last stream out through the gaps between the atoms, filling the entire Universe. If the time when everything in the entire visible Universe today was piled up in one place is set as 'time zero', this critical temperature (about 6000 K) was reached between 300,000 and 500,000 years later. This is the time of the first light, when matter and radiation are said to have 'decoupled'.

Since then, the Universe has expanded enormously, and the electromagnetic radiation that filled the Universe has been stretched and redshifted accordingly. It is straightforward to calculate the effect this has on the radiation. It turns light, just like the light from the Sun and stars, with a temperature of about 6000 K, into much longer wavelength radio waves, with a temperature of only a few Kelvin. It is exactly like the radiation in a microwave oven, but with a temperature of about *minus* 270 degrees Celsius.

This is exactly what Penzias and Wilson found, and which other radio astronomers soon confirmed: a hiss of cool microwave radiation, coming from all directions in space (from the gaps between the galaxies). This radiation has not interacted with anything at all from 500,000 years after time zero until it fell into the radio telescopes used to detect it.

Over the next two decades, many different radio telescopes, operating at many different wavelengths, identified the cosmic microwave background radiation and confirmed that it has exactly the right properties to be the highly redshifted first light from the Big Bang itself. These measurements also pinned down the temperature of this radiation very precisely (to 2.735 K). By 1978, the evidence that this really was the 'echo of the Big Bang' was so compelling that Penzias and Wilson received the Nobel Prize for their work.

A Smooth Beginning

Apart from the fact that it exists, however, the most important thing about the background radiation is that it is very smooth. To this accuracy (three decimal places, better than 0.01 per cent or 1 part in 10,000), the temperature is exactly the same from all parts of the sky. Matter and radiation were inextricably mixed up until the time of the first light. This meant that when matter and radiation decoupled, the hot gas that later went on to form stars and galaxies was also distributed very smoothly through the Universe, smooth to an accuracy of better than 0.01 per cent.

This tells us that the Big Bang itself was a very smooth event – and gives astronomers the task of explaining how things like stars and galaxies ever did get to clump together from such a smooth beginning.

The Birth of the Universe

In the ten years after Einstein came up with the general theory of relativity in 1916, mathematicians played with the equations, exploring the possibilities allowed by the laws of physics, because that's what they like to do – play with equations. It was only after Hubble and Humason discovered Hubble's Law – that redshift is proportional to distance – that anyone realized that the equations might be describing the actual expanding Universe in which we live. The person who first tried to use the equations to work out how the Universe was born was a Belgian astronomer by the name of Georges Lemaître. This was the beginning of Big Bang cosmology.

2

A TV screen when it is on, but not tuned in, will show white dots dancing on it. Some of these (about 1 per cent) are caused by cosmic background radiation.

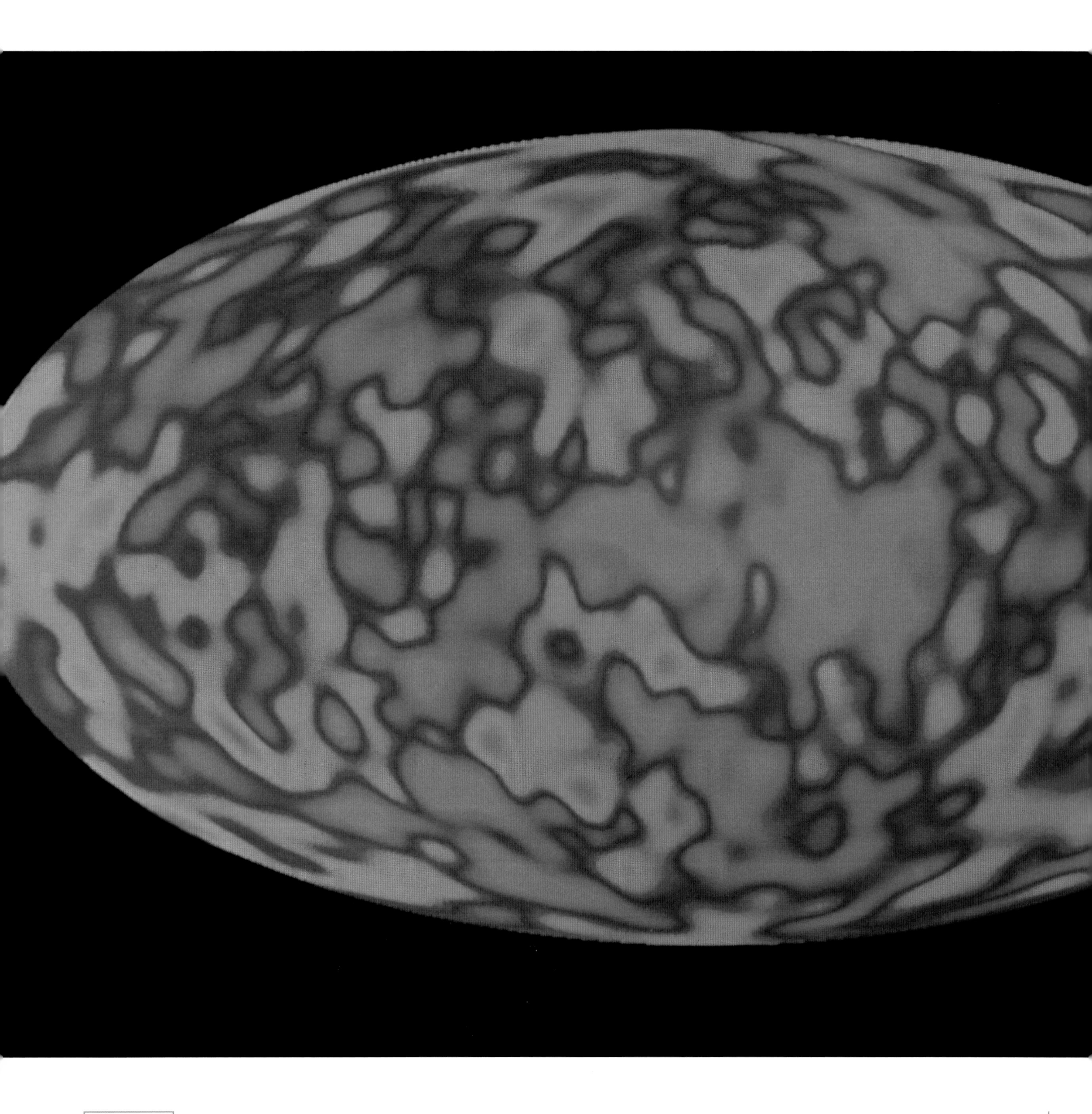

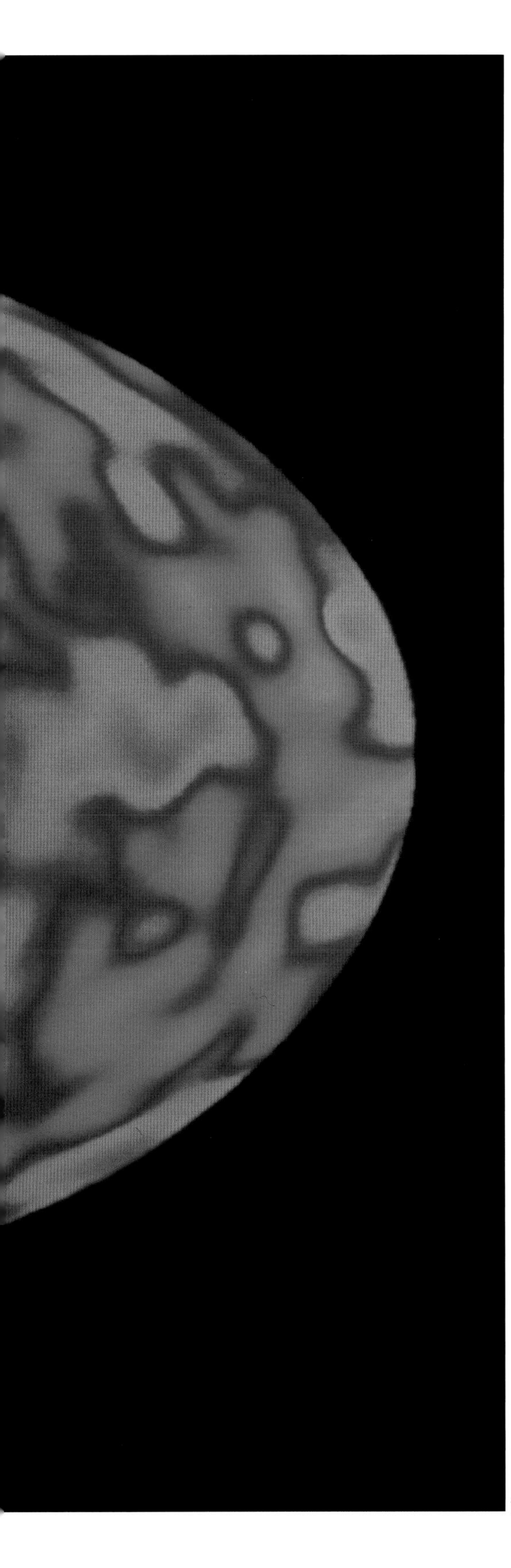

1. This false-colour image based on COBE data shows the whole sky in microwave 'light'. Pink shows hotter regions, and blue indicates cooler regions.

The Cosmic Egg

Lemaître developed his ideas about what we now call the Big Bang in the 1930s and 1940s. He realized that, because the Universe is expanding, with galaxies getting further apart, it must have been in a superdense state long ago, with everything in the visible Universe today packed together in a single lump of stuff with the same density as the nucleus of an atom (or a neutron star ▷ p. 71). He sometimes called this the 'Primal Atom', and sometimes referred to it as the 'Cosmic Egg'. The amazing thing about Lemaître's calculation is that the Cosmic Egg would only have been about 30 times the diameter of the Sun – much smaller than the Solar System. This gives you some idea of just how much empty space there is between the stars and galaxies; it also shows how much smaller than an atom an atomic nucleus is (which is why 'Primal Atom' is a bad choice of name – it should have been 'Primal Nucleus').

Lemaître didn't try to explain where the Cosmic Egg had come from. He had the idea that it was rather like a giant unstable atomic nucleus, and that it had 'split', or decayed, in just the same way that an unstable nucleus of an element such as uranium-235 splits into lighter elements. It is no coincidence that Lemaître developed these ideas at about the same time that physicists were first studying the natural decay of elements like uranium-235, and developing the first so-called atomic bombs. This depends on the process properly referred to as nuclear fission, but which is popularly known as 'splitting the atom' (so an 'atomic' bomb is really a nuclear bomb). This is why Lemaître, a great popularizer of science, referred to a Primal Atom.

This is an unfortunate analogy in one way, because it gives the impression that the Cosmic Egg was sitting somewhere in empty space for an indefinite time and then exploded outwards into space. Einstein's equations, however, tell us that there was no empty space for it to explode into. The egg contained all the matter in the Universe and all the empty space. The primal stuff, whatever it was, expanded outwards, because space itself expanded.

The Singular Beginning

Einstein's equations also tell us where Lemaître's Cosmic Egg came from. Although the density of an atomic nucleus, or a neutron star, is the most extreme density that matter can exist in today, if it is compressed even further it will shrink down into a point and become a black hole as it does so (▷p. 69). In 1965, the mathematician Roger Penrose proved, using Einstein's equations, that when a black hole forms, all the matter in it must fall into a single point inside the black hole, a point of infinite density and zero volume known, for obvious reasons, as a 'singularity'.

There is only one way in which it might be possible to avoid this happening, and that is if, as the matter falls towards the singularity, conditions become so extreme that the general theory of relativity is inadequate to describe what is going on. Almost certainly, this must happen before the density gets to infinity. But many tests have shown that the general theory of relativity is still a good description of what is going on until everything in the black hole has been squeezed into a tiny volume smaller than a sub-atomic particle, such as a proton or a neutron. This is good enough to take us beyond the Cosmic Egg and back to the

The radio telescope which discovered the background radiation had originally been designed and used for early experiments in sending TV signals across the Atlantic by satellite.

moment when time began.

The point is, that something collapsing inside a black hole towards a singularity looks a lot like something expanding outwards from a singularity, if time is reversed. It's easy to guess that the Cosmic Egg might have expanded out from a singularity, in a kind of mirror image of the way matter collapses inside black holes. But it is much harder to prove that this is the way the Universe works. Nevertheless, by 1970, Penrose (in collaboration with Stephen Hawking) had refined his earlier calculations to show that this is the case. The way the Universe is expanding today proves that it started out from a singularity – or at least, from a point so small and dense that the general theory of relativity breaks down, a point even smaller than a proton or a neutron. By the time it was as big as Lemaître's Primal Atom, the Universe was already expanding rapidly as space stretched, and the simplest description of the Universe we see around us is that it is the inside of a rapidly expanding black hole.

But there is a lot more to cosmology than this simple description. One of the most fertile areas of scientific research in the 1990s and 2000s is the developing understanding of how the very early stages in the expansion of the Universe have imprinted it with the structure we see around us today. The existence of clusters of galaxies can now be seen to match the predictions of our model of the very early Unniverse.

BLACK BODY RADIATION

Because a red-hot object is cooler than an object glowing orange, which is cooler than an object glowing blue-white, astronomers can tell which stars are hotter than others by looking at their colours. They can do even better than this, because there is a mathematical expression which describes very accurately how the spectrum of energy radiated by an object at different wavelengths depends on its temperature. This is called the 'black body curve' (detected by the Bell Laboratories Horn Antenna, left). It may seem odd to describe a glowing object as a 'black' body, but the name derives from the fact that the same mathematical expression describes how radiation is absorbed by a perfectly black object. So 'black body radiation' can come from very hot, bright objects.

With this meaning of the term, the radiation from the surface of the Sun corresponds to radiation from a black body with a temperature of 5800 K. This temperature is measured using the black body curve which matches the Sun's spectrum. In the same way, astronomers can measure the temperature of a star thousands of light years away by comparing its spectrum of light with the appropriate black body curve.

But black bodies need not be hot. The spectrum of microwave radiation coming from all directions in space, the cosmic background radiation, has a curve which corresponds almost perfectly (to an accuracy of about one part in a 100,000) with the radiation from a black body with a temperature of 2.7 K. This is the chilled, leftover radiation from the Big Bang itself.

1. Representation of a singularity in space time. The same representation could describe either a collapsing black hole or the expanding Universe.

2. As well as being an astronomer, the Belgian Georges Lemaître was also an ordained priest in the Catholic church.

THE FIRST FOUR MINUTES

Stellar nucleosynthesis describes how all the elements in the Universe except primordial hydrogen and helium were 'cooked' inside stars out of that hydrogen and helium (▷ p. 54). This understanding was essentially complete by the end of the 1950s. But one key puzzle remained: where did the original hydrogen and helium come from in the first place? The greatest triumph of cosmology in the 1960s was the explanation of how interactions taking place in the Big Bang itself produced the primordial elements, with exactly the proportion of hydrogen and helium that we see in the oldest stars today, in the span of just under four minutes.

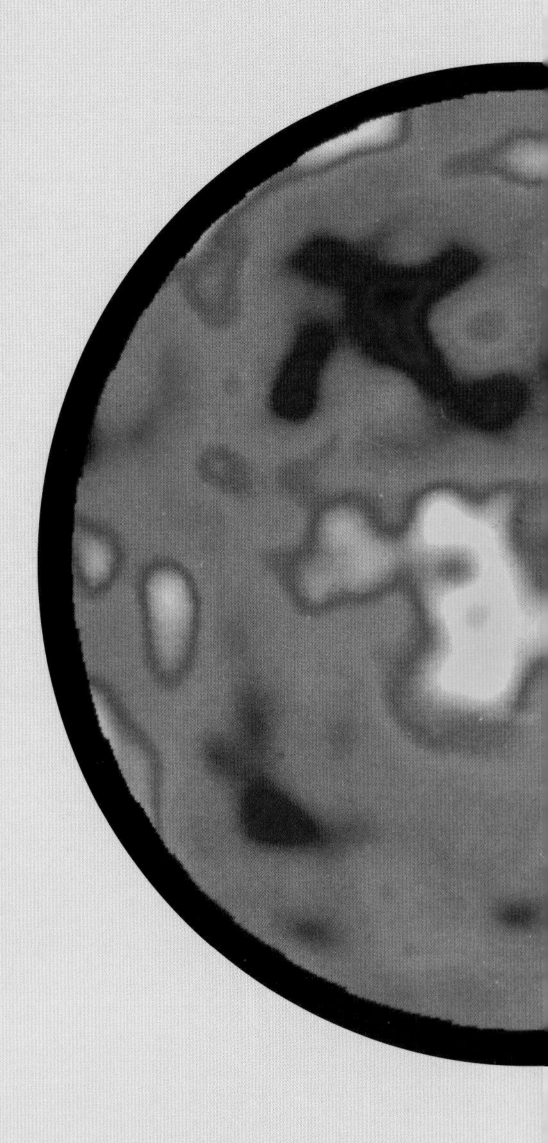

1. Fred Hoyle, whose insight led to an understanding of how the elements were made.

Matter from Radiation

The story of Big Bang nucleosynthesis begins just after the beginning of time. At time zero, the temperature everywhere was 100 billion Kelvin, so hot that most of the energy of the Universe was in the form of electromagnetic radiation, with protons and neutrons being created and destroyed by the radiation in line with Einstein's equation $E = mc^2$.

Matter began to 'freeze out' from the radiation when the temperature fell to 30 billion K, 0.1 s after time zero. At that time, the density of the entire Universe was 30 million times the density of water, but the energy was still mostly in the form of radiation. This radiation, however, could no longer create and destroy particles as easily as before (there was less E around, but the mc^2 needed to make each proton, say, was still the same). In the beginning, the radiation made equal numbers of protons and neutrons (plus lots of electrons), but because neutrons tend to decay, if they are not locked up in nuclei, spitting out an electron and turning into a proton, gradually the proportion of protons increased.

Freezing at Three Billion Degrees

By the time the temperature dropped to 3 billion K, 13.8 s after time zero, nuclei of deuterium (one proton and one neutron together) could form temporarily. They were soon knocked apart in collisions with other particles and conditions were at last starting to resemble what goes on inside stars today. Three minutes and two seconds after time

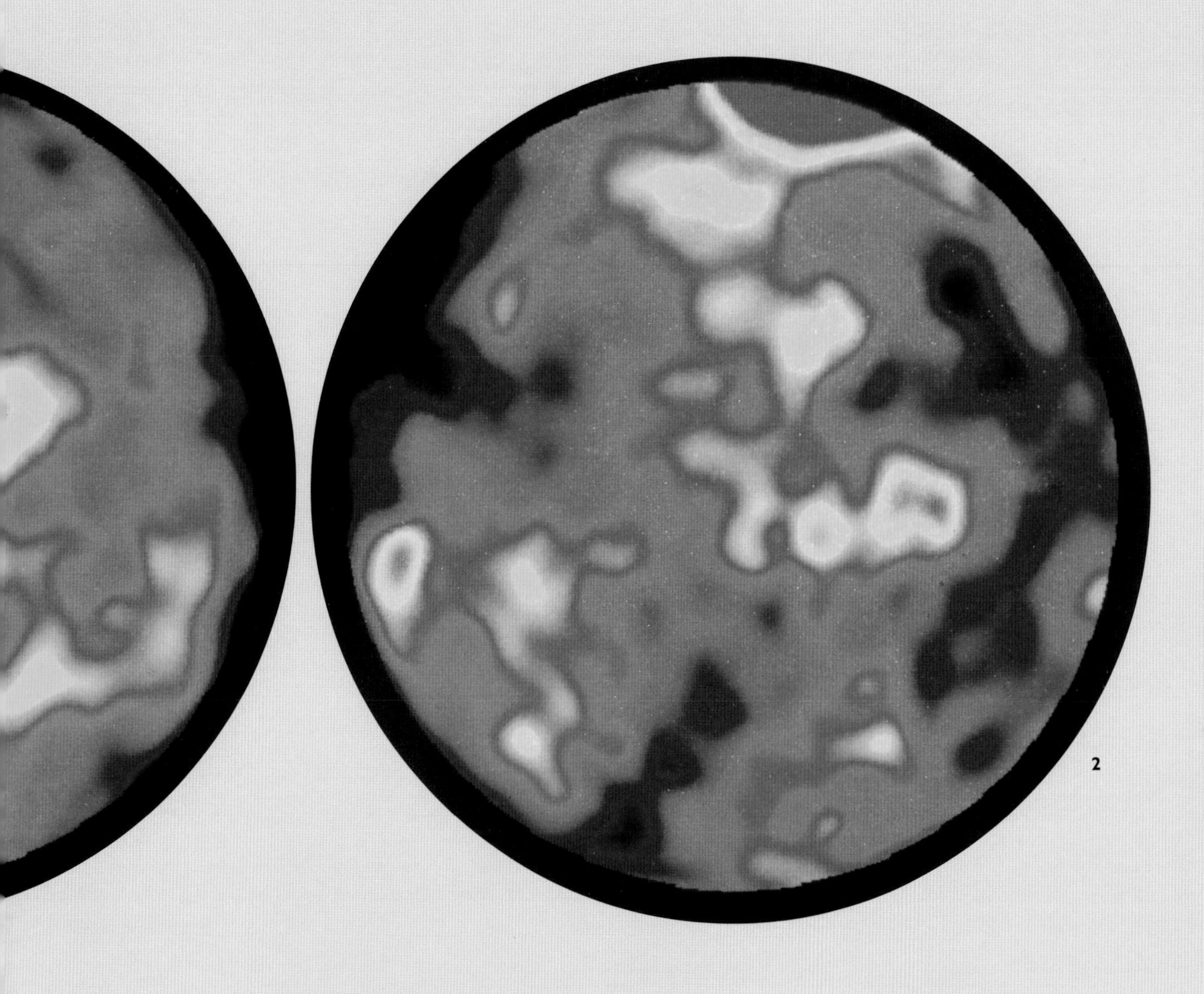

2. Detailed map of temperature variations when the Universe was half a million years old, based on data gathered by COBE over four years.

zero, the temperature had fallen to a billion K, just six times hotter than the heart of the Sun today, and the proportion of neutrons to protons had fallen to just 14 per cent. But they were saved from disappearing entirely because at last it was cool enough for nuclei of deuterium, and other light nuclei, to stick together permanently.

Saving Neutrons

In a flurry of nuclear reactions over the next few seconds, almost all the remaining neutrons in the Universe were locked up with protons in nuclei of helium-4, producing a mixture of about 25 per cent helium (in terms of mass), nearly 75 per cent hydrogen, and a trace of very light elements such as deuterium and lithium. This process of Big Bang nucleosynthesis finished about 3 minutes and 46 seconds after time zero – in round terms, four minutes after the beginning.

The standard Big Bang theory predicts exactly the proportion of these light elements actually seen in the oldest stars.

TOPIC LINKS

1.3 Stellar Evolution
p. 53 Cooking the Elements

2.2 Cosmology for Beginners
p. 94 The Steady State Model

2.4 The Accelerating Universe
p. 132 Quantum Fluctuations
pp. 134–5 Balloons and the Background Radiation

2.2 COSMOLOGY FOR BEGINNERS

Cosmology started out as a game played by mathematicians who liked to tinker with the equations of the general theory of relativity, which describes the behaviour of space, time and matter. The Universe is the biggest collection of space, time and matter they could think of. The game started soon after Albert Einstein discovered the general theory, in 1916. Before then, any views that people had about the origin of the Universe and its ultimate fate depended entirely on philosophy or religion, and had no scientific basis.

After Einstein's discovery, ideas about the evolution of the Universe abounded, based on his equations. But it was impossible to say which set of ideas matched the real Universe. It was only right at the end of the twentieth century that cosmologists were able to test their calculations with accurate measurements of the way the real Universe behaves.

A VARIETY OF UNIVERSES

Mathematicians can use the equations of the general theory to provide mathematical descriptions of different ways in which space and time interact in the presence of matter – that is, different ways in which a universe could change as time passes. Notice the use of the lower case 'u' on 'universe' there. Cosmologists reserve the capital, Universe, to refer to the real version of spacetime in which we live. They refer to their hypothetical mathematical versions as universes, or as 'models'. There are very many (perhaps an infinite number) of model universes that can be described using Einstein's equations, and the big trick is to find one that looks just like our own Universe.

Fortunately, our Universe turns out to be a very simple variation on the theme, and can be described using a very simple version of the equations. Einstein himself marvelled at this, saying, 'the most incomprehensible thing about the Universe is that it is comprehensible.'

But just how simple *is* the Universe?

Simple Models of Space and Time

The most basic division of cosmological models into categories can be described in terms of the expansion of the Universe. The Universe is expanding today but the gravity of all the stuff in the Universe is tugging on all the stuff in the Universe and trying to slow the expansion down.

The difference between the two main kinds of model can be pictured by imagining the difference between a baseball being hit and a rocket being launched. Even the most powerful baseball player in the world cannot hit the ball hard enough for it to escape from the pull of the Earth's gravity. No matter how high the ball

1

2

1 and 2. No baseball player can hit the ball so hard that it never falls back to Earth. You need a very powerful rocket to escape the Earth's gravitational pull.

flies, it will slow down, stop, and then fall back to Earth. But a sufficiently powerful rocket can launch at such a high speed that it can escape from the bonds imposed by the Earth's gravity, leave the vicinity of the Earth entirely, and never return. It is said to have achieved 'escape velocity' from Earth – a velocity which depends only on the mass of the Earth itself.

The question is, is the Universe expanding fast enough to escape from its own gravity – will it expand forever, or one day slow down and stop like the baseball and re-collapse? It is relatively easy to measure how fast it is expanding, but much harder to measure how much mass there is in the Universe, so it took a long time to find the answer.

Two Complications

This simple picture is not really quite so simple.

First, Einstein's equations also allow for the presence of a number, called the 'cosmological constant', which affects the expansion rate. It is given the Greek letter lambda (Λ) in those equations, but nothing in the equations tells us what value lambda has. Depending on its size, it could act as a kind of antigravity, making the Universe expand faster, or as an extra gravitational influence, slowing down the expansion. This is great fun for the mathematicians, because it gives them lots more models to play with. But studies of the expansion of the real Universe show that even if the lambda term does exist it must be very small, and until the 1990s astronomers usually set the constant as zero, to make life easier.

The other complication is more of an oddity than a complication. If you could throw a ball upwards at *exactly* the escape velocity, and nothing got in the way, it would keep going forever, but it would keep slowing down forever, and after a very long time it would seem to be hovering, high above the Earth (strictly speaking, infinitely far away), never to fall back. This curiosity is interesting, because the Universe itself seems to be very nearly in this state. It can be pictured another way, using geometry.

THE STEADY STATE MODEL

In the 1940s, three astronomers (Fred Hoyle, Tommy Gold and, right, Herman Bondi), came up with an idea to explain the expansion of the Universe without invoking a Big Bang. They pointed out that if new atoms of hydrogen were being created by expanding spacetime at a rate of just one new atom in every 10 billion cubic metres of space each year, enough atoms would be produced to make new galaxies to fill in the gaps as the old galaxies moved apart. At any moment in time (any cosmic epoch), the overall appearance would be much the same as it is in the Universe today. This became known as the 'Steady State' model.

This is a good example of how astronomers explore the Universe in their imagination, by thinking up models that might correspond to reality. When people argued that the continual creation of matter seemed rather an extravagant hypothesis, the supporters of the Steady State model pointed out that it didn't seem any more extravagant than creating all the matter all at once in a Big Bang.

The rivalry between the Big Bang and Steady State models encouraged astronomers to probe the Universe with radio and optical telescopes to find out which one was right. Eventually, this probing produced clear evidence that the Universe has changed as time has passed. It is not in a steady state. The discovery and investigation of the cosmic microwave background radiation then confirmed the reality of the Big Bang. So today we refer to the Big Bang theory (because it has passed the tests), whereas the Steady State is still just a model.

Einstein's Geometry

The general theory of relativity describes gravity in terms of bent spacetime. Around a black hole, the influence of the matter in the black hole bends spacetime round upon itself so that nothing can escape. The space around a black hole is like the surface of a sphere (or the surface of the Earth), and is said to be closed. Just as if you keep going in the same direction on Earth you go round the planet and back to where you started, so in closed space if you keep going in a straight line you go round the universe and back to where you started. The inside of a black hole really is a closed universe. Nothing can escape.

At the other extreme, gravity can bend space in the opposite sense. This is hard to picture, but an analogy would be the surface of a saddle, or a mountain pass, which curves away in all directions. Such a surface is said to be open. A closed universe is one which cannot escape from its own gravitational grip; an open universe is one which is expanding faster than its own escape velocity.

But there is a special case in which space is flat, like the surface of a smooth desktop. In Einstein's geometry, this is the equivalent of the special case where the ball is travelling upwards from the Earth at exactly the escape velocity. It is the only unique kind of geometry allowed by the general theory of relativity – there are lots of open universes, and lots of closed universes, but only one flat universe. And the real Universe seems to have a geometry indistinguishable from this very special case.

Close to Critical

Because the flat universe model is unique, cosmologists use it as a benchmark against which to measure other models. The flat universe model is said to have 'critical density', which means that the density of the universe is exactly right to make space flat. Cosmologists measure the density of the Universe (or model universes) in terms of a parameter called omega (Ω). It is also known as the flatness parameter. For a flat universe $\Omega = 1$. For an open universe, Ω is less than 1, and for a closed universe, Ω is bigger than 1.

The Universe we live in is expanding, and that means that the density is decreasing as time passes. This affects the value of Ω at any moment of cosmic time (any epoch), but it does not affect which side of the critical dividing line Ω sits. Leaving aside some of the more exotic complications that can be caused by a cosmological constant, if the Universe was dense enough to be closed when it

1. In an open universe, the angles of a triangle add up to less than 180°.

2. In a flat universe, the angles of a triangle add up to exactly 180°.

3. In a closed universe, the angles of a triangle add up to more than 180°.

Alexander Friedmann's application of the general theory of relativity produced a variety of model universes in the early 1920s – before astronomers knew for sure that there was a Universe beyond the Milky Way.

1

emerged from the Big Bang, then it will stay dense enough to be closed. And if it started out open, then it will always be open.

Going to Extremes

Whichever way the Universe started out, the fact is that it shifts further and further from that critical density as time passes. If it started out with Ω just less than 1, as it expands and the density decreases, the value of Ω gets less and less. If i t started out with Ω just bigger than 1, the expansion proceeds more slowly and the value of Ω gets bigger and bigger as time passes. By now, some 14 billion years after the Big Bang, there has been ample time for this effect to have been at work. The Universe should have evolved to one extreme or the other. But when we look at the real Universe, we see that it is still very close to the critical density.

It is hard to measure the density of the entire Universe, but astronomers can make a good stab at it. First, they count all the bright galaxies in a chosen volume of space, and estimate the mass of all the bright stars in those galaxies. Then, from the way galaxies move in clusters under the influence of gravity they can estimate how much 'dark stuff', or 'dark matter'(▷ p. 117), there is, tugging on the bright stuff gravitationally, for every galaxy they see. Adding all this up, they find that there is matter to account for at least 10 percent of the critical density of matter around, and probably at least 30 percent. So, based on the dynamics of galaxies, Ω is at least 0.1 and may be bigger than 0.3. But for 14 billion years, Ω has been getting further and further away from 1. For Ω to be as big as 0.1 today means that in the first second of the Big Bang it must have been within one part in 10^{60} (a 1 followed by 60 zeroes) of being precisely 1. This means that the Universe was born flat to an accuracy of 1 part in 10^{60}. The flatness of the Universe in the Big Bang is the most accurately determined parameter in the whole of science.

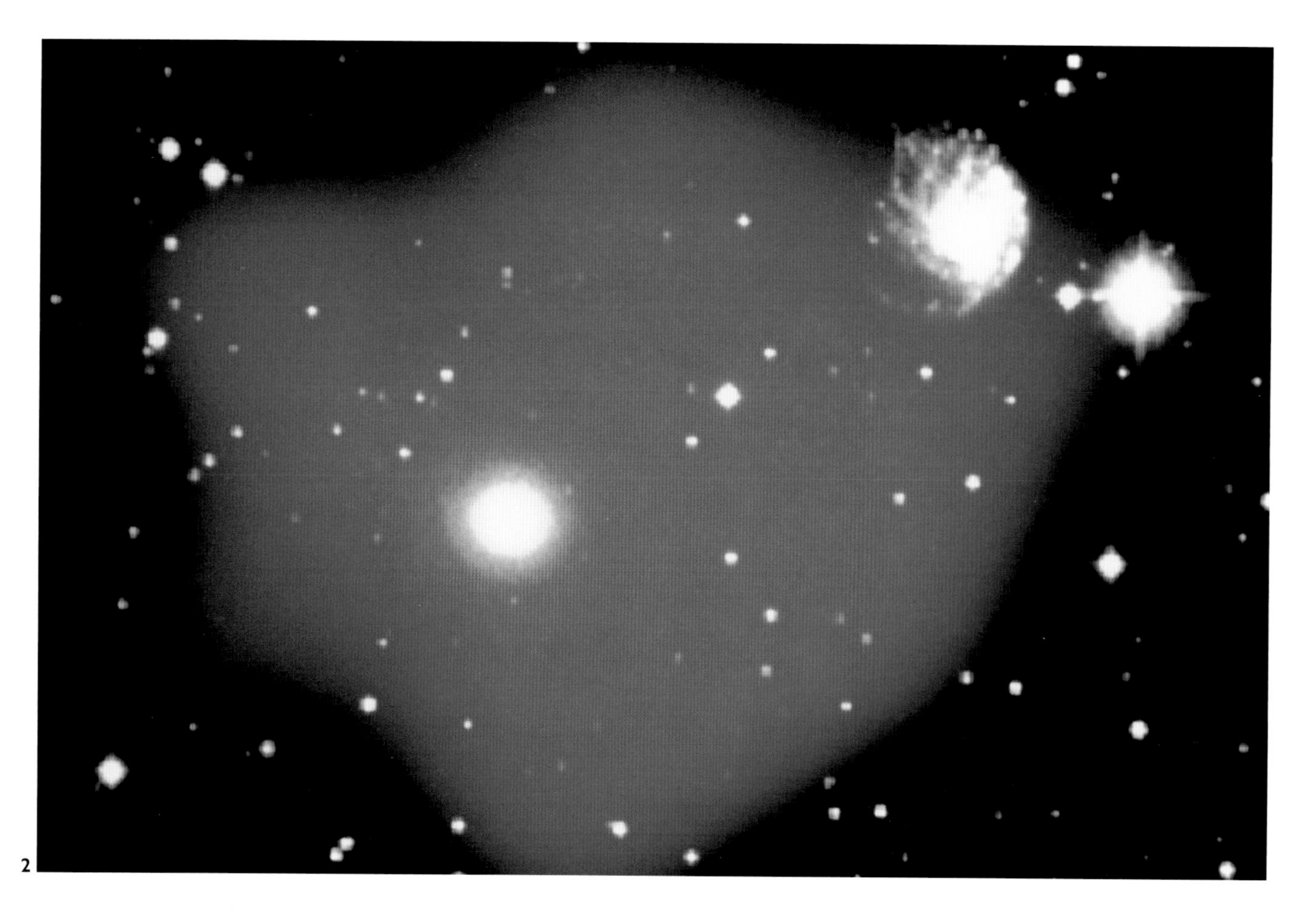

1. True colour image of the central region of the Fornax cluster of galaxies about 70 million light years away from us.

2. Evidence for dark matter. In this composite image a cloud of gas detected only at x-ray wavelengths (coloured magenta here) fills the space between galaxies in the cluster NGC2300.

More than Meets the Eye

Many cosmologists believe that this must mean that the value of the flatness parameter Ω in the Big Bang was precisely 1. Because the critical density is the only special density there is in cosmology, it is hard to see how the Universe could have started out so close to critical without actually being at critical density. The fact that the amount of matter we can locate today only corresponds to a value of between 0.1 and 0.3 is not a problem, they say, because our observations of the Universe are incomplete. We cannot claim to have discovered everything in the Universe, so there must be more dark stuff, of one kind or another, which makes Ω bigger.

The important point is that a value for Ω of 1 is the only value that stays the same as the Universe expands. If Ω is 1 in the beginning then it is always 1. As the Universe expands and slows down, the density decreases at exactly the right rate so that at each epoch the speed with which the Universe is expanding is still exactly the escape velocity from itself. A few cosmologists, however, hope that there is even more mass than this.

The Phoenix Universe

Where did the Big Bang come from in the first place? Until the end of the twentieth century, one attractive idea proposed by the mathematical cosmologists was a model universe in which there was an endless cycle of birth, death and rebirth. Like the mythical Phoenix, which lays a single egg, dies, and then

The average density of the Universe is equivalent to between ten and 100 hydrogen atoms in every cubic kilometre of space.

1

emerges from the egg when it is heated in a fire, such a universe recreates itself. It does so in a fireball like the Big Bang. The mathematics still hold up – a universe (or universes) like this *could* exist according to the known laws of physics. But could this model actually represent the real Universe? One thing it requires is that the density of the universe is higher (preferably a lot higher) than the critical density needed for flatness.

If Ω is bigger than 1, sooner or later the expansion of the universe will stop and then reverse, so that eventually everything in the universe will plunge together into what is sometimes known as the 'Big Crunch' (occasionally, and more pompously, as the 'Omega Point'). In the early stages of the collapse, life could go on pretty much as it does now, but the light from distant galaxies would be blueshifted, not redshifted, as space began to shrink, rather than expand. As time passed, the background radiation would get hotter, but it would take a long time for this to begin to cause problems.

Crunch and Bang

The first real change in the overall appearance of the shrinking universe occurs when it has shrunk to about 1 per cent of the size of our Universe today and the galaxies begin to merge with one another (this reflects the fact that in very round terms the distances between galaxies today are 100 times larger than the sizes of galaxies themselves). Even then, the temperature of

THE QUANTUM OF TIME

Cosmologists know that actual singularities (points of infinite density and zero volume – ▷p. 89) cannot really exist inside black holes and that the Universe as we know it cannot have been born out of a literal singularity. This is because space and time themselves are 'quantized' – there is a smallest possible length and a smallest possible time, both specified by the laws of quantum physics. The smallest length (called the 'Planck length', in honour of the quantum pioneer Max Planck, right) is 10^{-33} cm, one hundredth of a billionth of a billionth the size of a proton. The smallest possible time (called the 'Planck time') is the time it would take light to cross this tiny distance, which is 10^{-43}s.

Although these are unimaginably small quantities, they are not zero. This is important, because it means that physicists do not have to divide by zero in their equations, so they do not get answers of infinity out of their calculations (for example, when they work out the density of the Universe in the beginning). The implication is that, whatever the event was that created our Universe, the Universe was 'born' with a finite density at an age of 10^{-43}s, and that this was the moment when time began. In a collapsing universe heading for a Big Crunch (see opposite), the 'bounce' that turns the universe inside out would happen not at a singularity, but when the collapse was still 10^{-43}s away from reaching the singularity.

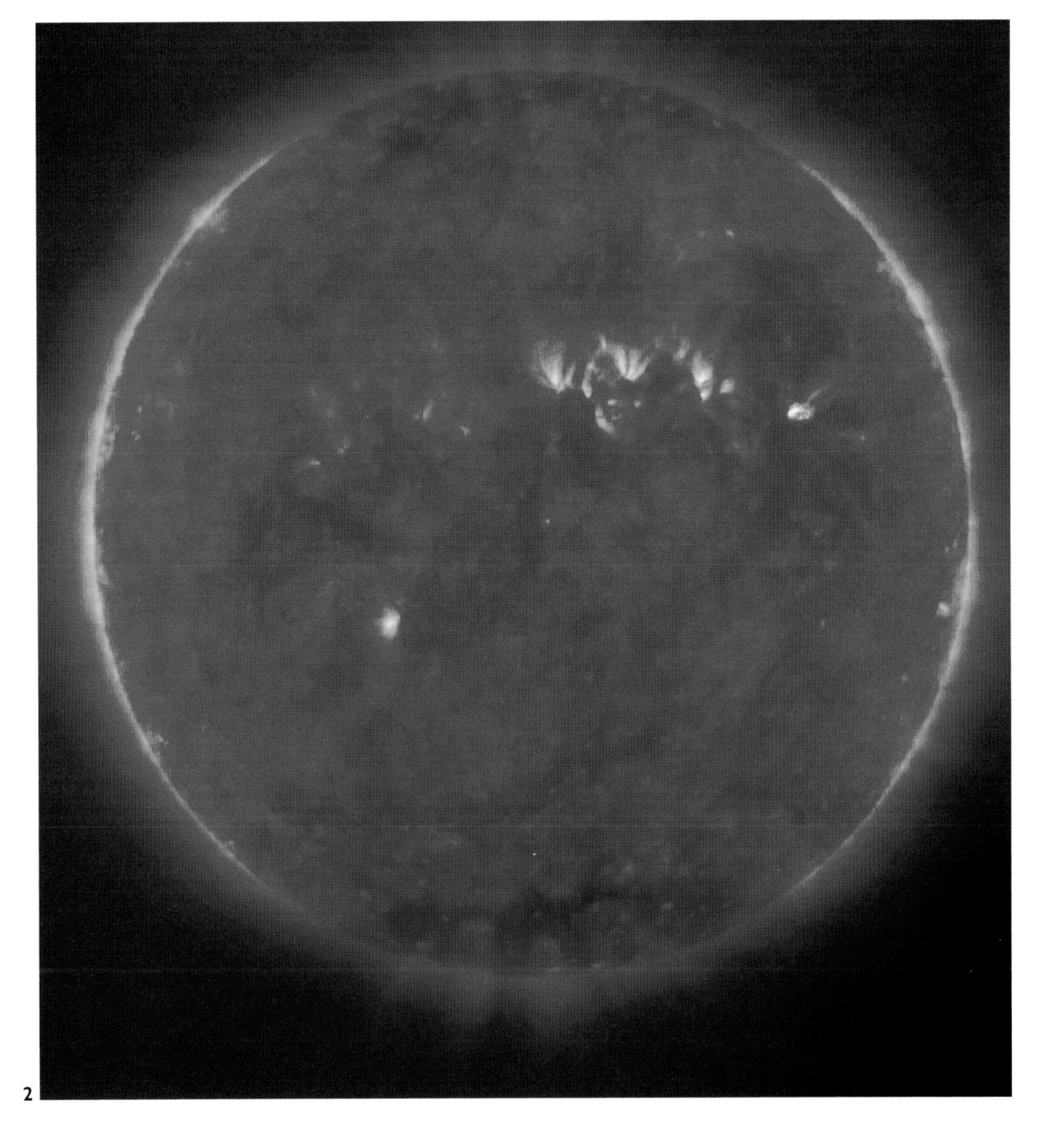

2

1. Like the mythical Phoenix, the universe may not only have been born in fire, but had probably also experienced previous cycles of life and death.

2. In the late stages of a collapsing universe, the blue-shifted light from the whole sky would be as hot as the Sun.

the background radiation would only be about 100 K, and it is entirely possible that life could still exist under those circumstances.

By the time the universe has shrunk to about 0.1 per cent of the size of our Universe now, the blueshifting of the background radiation would make the entire sky as bright as the surface of the Sun is today. The 'background' radiation would have a temperature of thousands of Kelvin and life as we know it would be impossible. Soon after, about a year before the Big Crunch, the background radiation would be hotter than the inside of a star and all life would be extinct. Stars themselves would be torn apart by the radiation and broken down into their constituent particles. An hour away from the Big Crunch, based on the rate at which spacetime is contracting then, the huge black holes that exist at the centres of galaxies would begin to merge with one another. But

Alexander Friedmann, who first applied Einstein's equations to a variety of model universes, died of pneumonia after he caught a chill while flying in a balloon to study the weather.

the merging of black holes is an all-or-nothing affair, and instead of the rest of the collapse taking an hour it would happen instantly, as the singularities inside the black holes came together, causing the Big Crunch itself.

What would happen then is largely guesswork. But the most popular guess is that this dramatic mingling of singularities would cause a 'bounce', turning the collapse inside out and making the universe expand in a new Big Bang. According to this picture, our Big Bang may have been the Big Crunch of a previous cycle of expansion and collapse, in an endlessly repeating rhythm.

Reversing Time

There is one extremely curious feature of a collapsing universe – at least, according to some of the mathematicians who explore these possibilities. In the 1960s, the cosmologist Tommy Gold (one of the co-founders of the Steady State model) suggested that during the collapsing phase time might run backwards.

Gold's original speculation was not based on detailed mathematics, but is still intriguing. Because the laws of physics have no inbuilt 'arrow of time', they would work equally well if the direction of time in the equations were reversed. Gold said that, according to those laws, stars in a collapsing universe would soak up radiation from space instead of emitting light. Inside the stars, the incoming energy would drive nuclear reactions that converted helium into hydrogen, while on a planet like our own, ice cubes would radiate away heat and grow larger, while living things 'grew' from old age to youth.

With time running backwards, you might think that the world would look very bizarre. But the crucial twist in Gold's story is that inhabitants of such a world might never notice. If time is running backwards, the thought processes that make intelligent beings intelligent will also run backwards. Any intelligent being living in the contracting phase of the universe would think 'backwards' compared with us, and would still 'see' heat flowing from hotter objects and into cooler objects and so on. The punch line is that we could be living in a contracting universe and are totally unaware of it!

This intriguing, but infuriating, suggestion has roused several cosmologists, including Paul Davies and Stephen Hawking, to try to find a mathematical description for what goes on in such a universe and prove whether or not time can run backwards. They have not yet succeeded in finding a definite answer to the question of whether time runs backwards when a universe contracts, and you can get some idea of how difficult a problem this is from the fact that Hawking has changed his mind at least twice on the answer to this puzzle. It is probably just as well for the sanity of the mathematical cosmologists that it now seems as if the question will remain a hypothetical one. There may indeed just be enough stuff in the Universe we live in to make it flat, but there is now compelling evidence that there is no more mass-energy than this, and that the Universe will expand forever. One of the big puzzles in cosmology is finding enough mass to flatten the Universe.

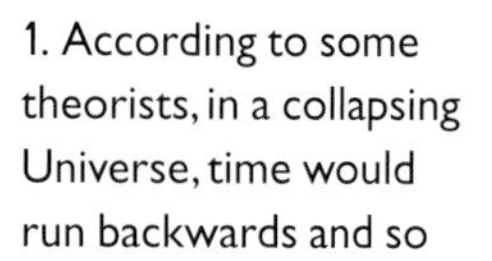

1. According to some theorists, in a collapsing Universe, time would run backwards and so drops of water would fall upwards from the ripples in a pond.

THE ARROW OF TIME

One of the most puzzling things in science is the origin of the 'arrow of time'. We all know that there is a difference between the past and the future, but where does this difference come from? There is nothing in the laws of physics which makes the distinction. In the classic example of two billiard balls colliding and moving apart, the laws of physics 'work' just as well to describe the same collision with time running backwards. However, the arrow of time seems to have something to do with the way very large numbers of things interact with one another. In the break of snooker, for example, it is quite clear in which direction time goes.

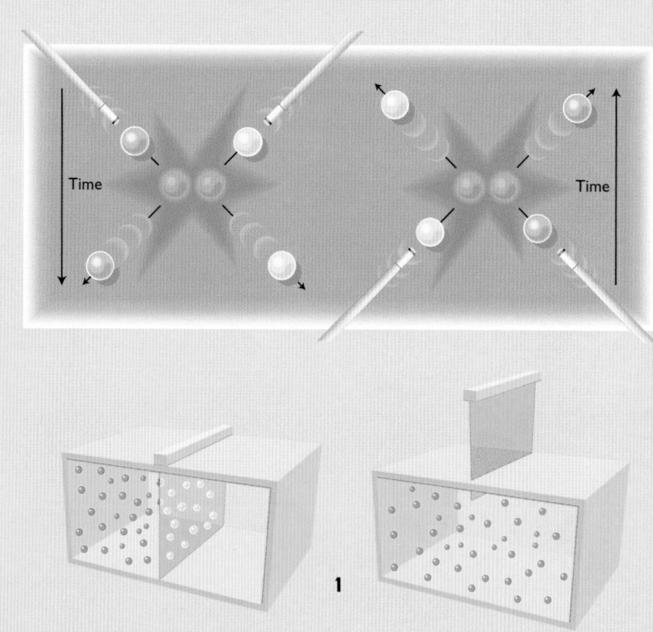

1

Time in a Box

Think of a simple box, divided in two by a sliding partition, with smoke in one half and empty space in the other. When you slide open the partition, the smoke spreads out to fill up the box. However, no matter how long you wait, you will never see the smoke move back into one half of the box so that you can slide back the partition and trap it. If you saw two pictures of the box, one with smoke in one half of the box and one with smoke in all of the box, you would know which picture was taken first.

Mathematically, the difference between the two situations is that you need less information to describe the full box. There is a pattern (or order) in the half-empty box (half-empty, half-full) that is lost when the box is uniformly full of smoke. Information, or order, is measured by a quantity called entropy, in such a way that a decrease in information (increase in disorder) corresponds to an increase in entropy. Overall, in the Universe at large, entropy always increases as time passes.

2

3

Local Order

Left to their own devices, things wear out. Cars rust, glasses break, houses fall down. Disorder increases. We can only make order locally (making cars, building houses and so on) using energy. On Earth, the energy comes, ultimately, from sunlight. But the increase in entropy, caused by the processes inside the Sun which release energy, is much bigger than the decrease in entropy caused by the action of life on Earth. In the Universe at large, entropy always increases, even though it may temporarily decrease on a planet like Earth. Time points in the direction of increasing entropy.

1. Although individual collisions between atoms and molecules seem to be reversible, the behaviour of lots of molecules in a gas reveals the direction of time.

2 and 3. Rusting cars (left) and burning fires (4 and 5 above) both indicate the arrow of time in the real world.

Time and the Universe

The Universe was born in a state of low entropy. Processes going on inside the Sun and stars increase the entropy of the Universe as energy pours out into the cold of space.

Another indication of the arrow of time is that in nature heat always flows from a hotter object to a cooler object. So another way of defining the arrow of time is that it points away from the hot Big Bang and into a cold future. When all the stars have burned out, everything in the Universe will be at the same temperature. It will be uniform, with no pattern or order, and there will be no way of telling one locality from another. Time will come to an end. This is called the 'heat death' of the Universe and it is inevitable if the Universe is destined to expand forever. The best evidence today is that this will be the fate of our Universe.

TOPIC LINKS

2.1 The Big Bang
p. 87 The Singular Beginning

2.2 Cosmology for beginners
p. 100 Reversing Time

2.4 The Accelerating Universe
p. 129 The Fate of the Universe

4.2 A Choice of Universes
p. 209 Hawking's Universe

2.3 MISSING MASS AND THE BIRTH OF TIME

After the 1960s, cosmologists began to accept that the Big Bang idea was more than just a model and offered a good description of the real Universe. But as it stood at the time, this was far from being an exact description of the Universe. The basic idea of the Big Bang seemed sound, but there were problems relating the detailed appearance of the Universe we live in to the physics of the Big Bang itself. These puzzles centre around the observational evidence that the Universe we live in is uniform but nevertheless contains irregularities.

A breakthrough came when particle physicists started to apply their ideas, based on studies of what happens to particles at very high energies, to the physics of the Big Bang. The resolution of those puzzles involved new ideas about what happened in the earliest moments of the existence of the Universe, and new observations to test those ideas. The result was a golden age of cosmology, which is still going on today.

Previous page. Optical image of the Omega Nebula 6000 light years away from Earth in the constellation Sagittarius.

BIG BANG PROBLEMS

Before about 1960, cosmology was still a mathematical game, in which a few experts (probably no more than 20 in the whole world) explored the possible model universes allowed by the equations of the general theory of relativity. In the 1960s, they were delighted, but rather surprised, to find that one of those models, the Big Bang, actually seemed to provide a very good description of the Universe we live in. The discovery of the background cosmic radiation, and the elucidation of how the primordial hydrogen and helium that went into the first generation of stars in the first four minutes of the Big Bang, made people start to take the Big Bang model seriously. But in the 1970s, cosmologists (by now there were hundreds of them) began to realize that there was a problem they had never expected with the Big Bang theory: it was, in a sense, too good to be true.

1

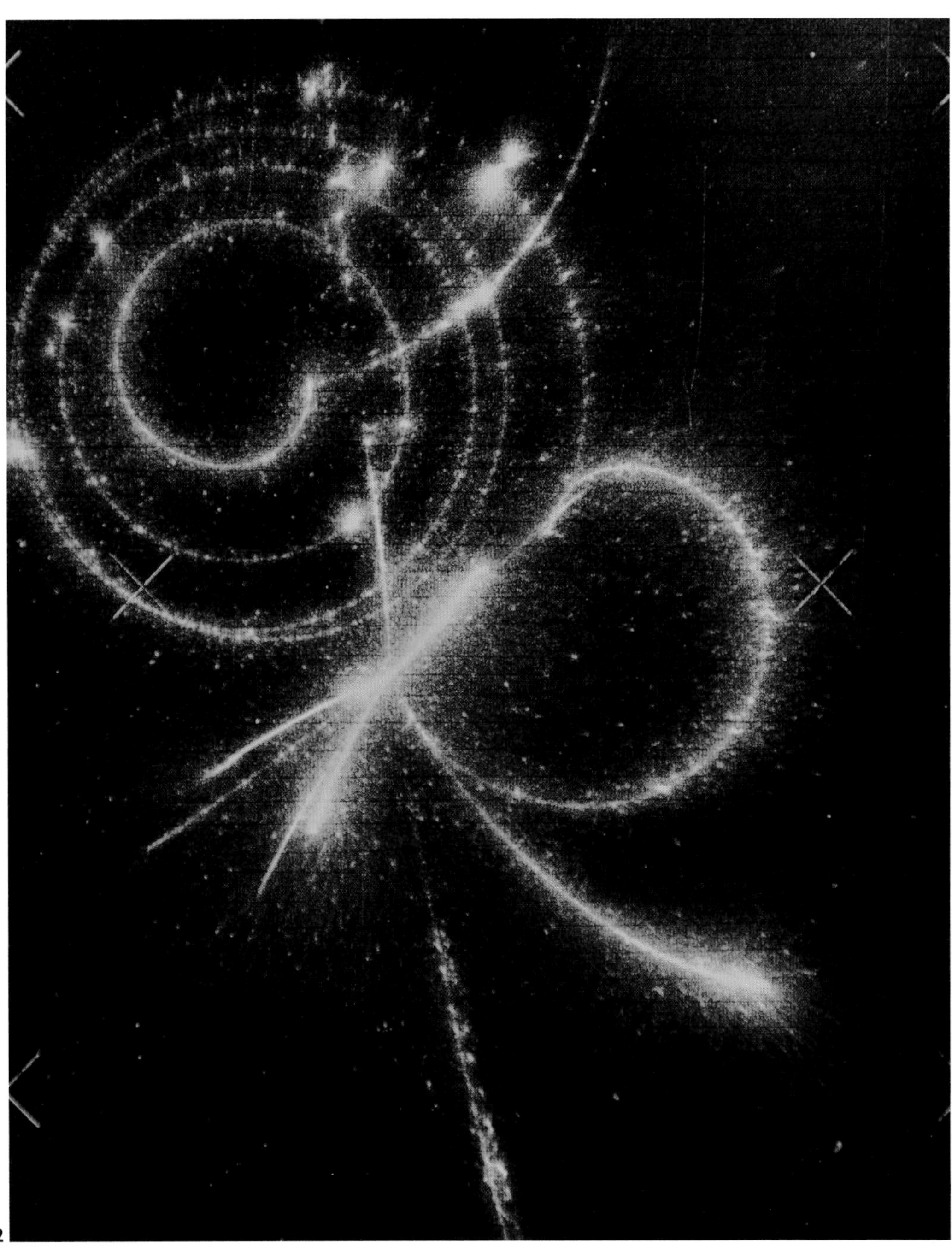

2

1 and 2. Photographs showing the interactions between sub-atomic particles in experiments at the high energy accelerators of CERN at Geneva.

It took only 0.000000000001 s (10^{-12}s) for what was to become the entire visible Universe to expand from a size smaller than an atom to the size of the Solar System.

1. On the scale of stars and galaxies, the universe doesn't look smooth. Stars form patterns like the southern cross.

2. Flatness can be a matter of perspective – the ground beneath our feet seems flat even though the Earth is round.

The Flatness Problem

In fact, there were several 'too good to be true' problems that puzzled cosmologists in the 1970s. The first is the question of why the Universe is so nearly (perhaps precisely) flat (▷ p. 95). As we have seen, whichever side of the critical density it started out from in the Big Bang, the Universe should have got further and further away from $\Omega = 1$ as time passed. For the Universe still to be so nearly flat after nearly 15 billion years is about as likely as balancing a pencil very carefully on its point on a table, then going away for 15 billion years and coming back to find that the pencil is still as you left it. The only explanation seemed to be that there ought to be a law of nature which forced the Universe to be precisely flat. But, in the 1970s, nobody could think what that law might be.

The Smoothness Problem

Another big puzzle about the Universe is that it is incredibly smooth. You might not think so, looking at the patterns made by stars on the night sky, or the way galaxies are grouped in clusters across the dark expanse of the Universe. But these are really small-scale effects. What matters is the smoothness of the dark space itself – in Einstein's language, the smoothness of spacetime. From a distance, the surface of the Earth looks smooth and, compared with its diameter, it is. A mountain 12.7 km high is only a pimple corresponding to one-thousandth of the diameter of the planet, even though it looks enormous to us. In the same way, the uniformity of the Universe has to be measured on the largest possible scale, and that means using the cosmic background radiation – the echo of the Big Bang.

The background radiation has the same temperature from all parts of the sky – it is

2

Although nothing can travel through space faster than light, space itself can expand faster than light.

1

2

isotropic. This shows that the Universe emerged from the fireball of the Big Bang in a very smooth state. We have to get used to the idea that even clusters of galaxies are like tiny pimples on the smoothness of the Universe.

There's another way of looking at this. The Universe isn't just the same in all directions, it is the same (on average) *everywhere*, once you allow for the effects of expansion. The pimples are distributed at random across the face of space. The buzzword this time is 'homogeneity'. You can see the homogeneity of space by looking at two stunning pictures from the Hubble Space Telescope, called Deep Field North and Deep Field South. These are images of very distant galaxies seen on opposite sides of the sky. But apart from superficial differences, they look exactly the same. The same kinds of galaxies are clustered together in the same kind of way. There is nothing in the pictures to tell you which part of the Universe those galaxies inhabit, although they occupy regions of space billions of light years apart (just as there is nothing in a picture of a pimple to tell you which cheek of an acne sufferer it afflicts). There are no 'lumps' in the Universe, except for the galaxies and clusters of galaxies themselves. It is homogeneous. Or nearly; there are, after all, galaxies. And in the 1970s, this was another problem.

The Problem of Galaxies

As observations of the background radiation got better and its smoothness was measured with increasing precision, it began to be difficult to see how galaxies could have formed at all. Galaxies start out when huge clouds of gas collapse under their own gravitational pull, their own weight. But this collapse can only

3

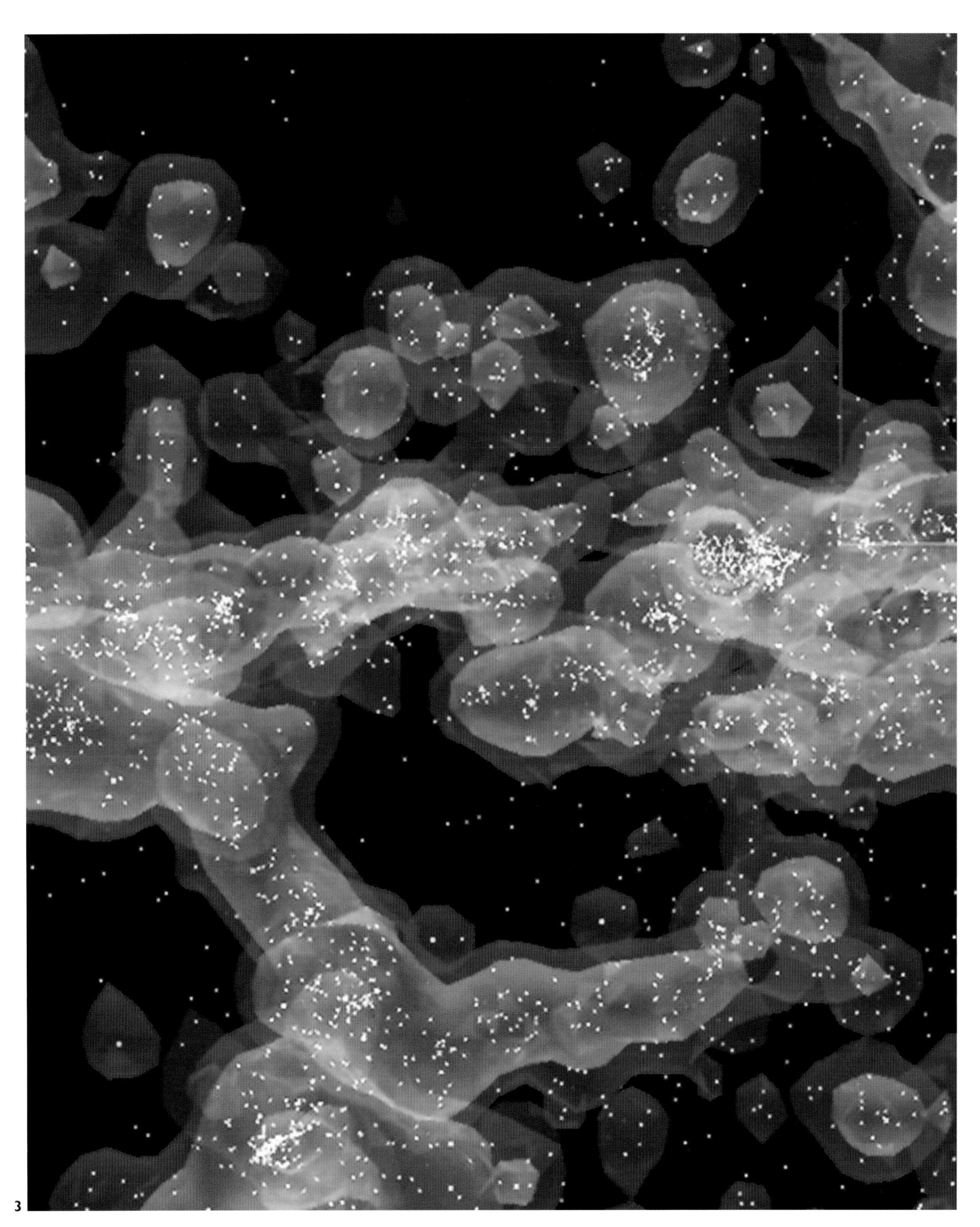

1 and 2. The universe looks the same in all directions. The top image is a picture obtained by the Hubble Space Telescope looking north on the sky. The bottom picture is a similar image obtained looking south.

3. The distribution of galaxies on the sky is not entirely random, but forms clumps and filaments.

1. The microwaves used to map the Universe are similar to the ones used in telecommunications.

2. Alan Guth, one of the pioneers of the theory of inflation.

3. (opposite) Computer simulation of the way matter clumps together in the expanding Universe.

begin if some regions of the Universe are more dense than others. If everything were perfectly smooth, gravity would pull evenly in all directions, and everything would get carried along with the expansion of the Universe. Like seeding clouds to make rain, you need some initial lumpiness to provide the seeds on which galaxies can grow. These seeds must have been present at the end of the fireball era of the Big Bang, 300,000–500,000 years after the moment when time began, at the time when the radiation that became the microwave background last interacted with matter.

So the unevenness in the Universe from that time, that led to the formation of galaxies, should have left its imprint on the microwave radiation. Why hadn't this been seen?

The answer was that it was too small. When cosmologists carried the calculations through, they found that the seeds from which galaxies and clusters of galaxies formed should indeed have left a mark on the background radiation, but a mark corresponding to differences in temperature today from one part of the sky to another of only about 30 millionths of a degree. Nothing highlights more clearly just how insignificant galaxies (let alone stars, planets and people) are to the cosmos at large. There was no hope (at that time) of detecting, from the ground, such tiny ripples in the background radiation. In the mid-1970s, a group of astronomers took up the challenge of designing and building a satellite which could look for the predicted disturbances in the background radiation. The result, COBE, did not fly until the end of the 1980s, when it triumphantly confirmed the predictions of the theorists. But by then, the theory of the Universe had itself undergone a radical rethinking that solved the problems of the Big Bang.

Flattening the Universe

The standard model of the Big Bang describes everything that happened in the Universe from about 0.0001 s (10^{-4}s) after 'time zero', when the temperature of the Universe was 1000 billion degrees (10^{12} K), up to the time about half a million years later when matter and radiation decoupled at a temperature of 6000 K. It never pretended to explain what went before that, when temperatures were higher than 10^{12} K. The understanding of physics that existed in the 1960s was inadequate to describe what went on that close to the singularity. But as physicists on Earth

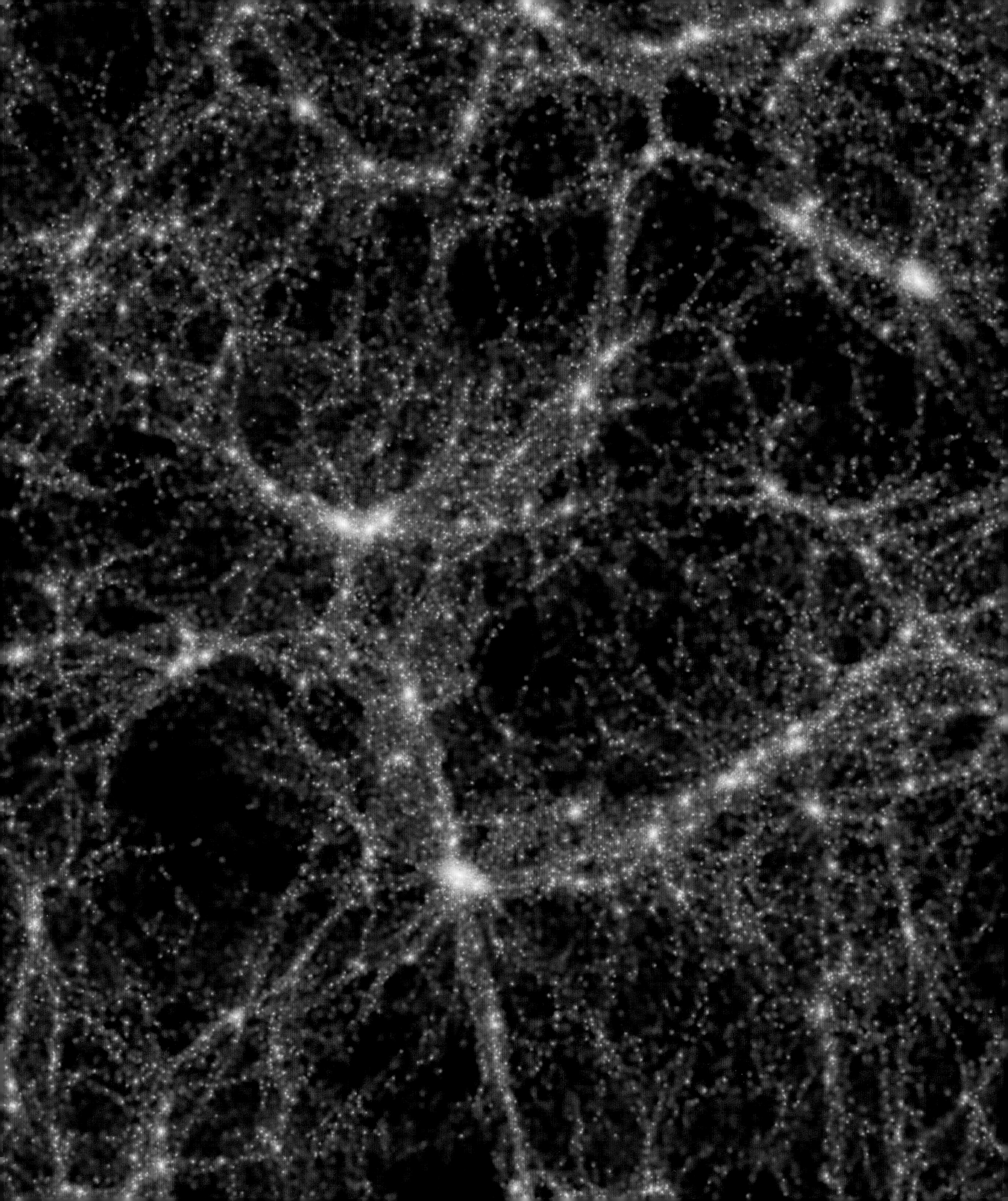

began to explore such extreme conditions, both with their particle accelerators and with new mathematical models and theories, the new discoveries offered a way to explain how the Universe got to be so flat and smooth.

Two New Beginnings

As is usually the case in science, progress in solving the puzzles was made because the time was ripe and both new technology and new ways of thinking about the puzzles became available. It is very rare that progress happens because of the unique insight of a lone genius.

In the late 1970s, cosmology was transformed by an influx of ideas from particle physicists, used to thinking about what goes on at the enormously high energies corresponding to temperatures above 10^{12} K. Two people, on opposite sides of the world, independently hit on the same solution to the puzzles of the Big Bang. This was not a result of a strange coincidence, but because the answer was built into the laws of particle physics that had been unveiled in the 1970s, and because their training in particle physics gave these men exactly the background needed to help solve those puzzles.

The two young researchers who made the breakthrough were Andrei Linde (then working in Moscow, but later in the United States), and Alan Guth (then working at the Stanford Linear Accelerator Center in California). Their ideas involve some very deep physics, but, even without going into the technical details, it is straightforward to see how they solved the problems of the Big Bang.

Universal Inflation

All of the problems of the Big Bang could be solved if the Universe had expanded enormously in the first split-second after time zero, long before the beginning of the standard Big Bang model. In this scenario, during that first split-second, something took

BENDING LIGHT

One of the first tests of the general theory of relativity measured the way light from distant stars was bent as it passed close by the Sun during the total solar eclipse of 1919. Now, almost 100 years later, astronomers can study the way light from far across the Universe is bent by whole galaxies (or even clusters of galaxies, right) that lie between us and the source of the light.

In the mid-1980s, astronomers discovered two huge, luminous arcs on the sky (they are huge in reality, but look small because they are far away). Each is about 90 000 parsecs long and spanning about 9000 parsecs. More arcs have since been discovered. In the best example, the light from the arc has the same spectrum and a redshift of 0.724, but the arc is along the same line of sight as a cluster of galaxies at a redshift of 0.374. The implication is that the light in the arc comes from an object (almost certainly a galaxy) about twice as far away as the cluster. This light has been bent by the gravity of the cluster in its way to us, forming a magnified and distorted image of the galaxy which produced it.

Calculations using Einstein's general theory show that in all of the arcs investigated so far, the galaxy clusters involved in this lensing process cannot produce such high magnifications unless they contain at least ten times as much mass as we can see in the form of bright stars in the component galaxies. This is powerful independent evidence to support the idea that at least 90 per cent of the matter in the Universe is dark stuff.

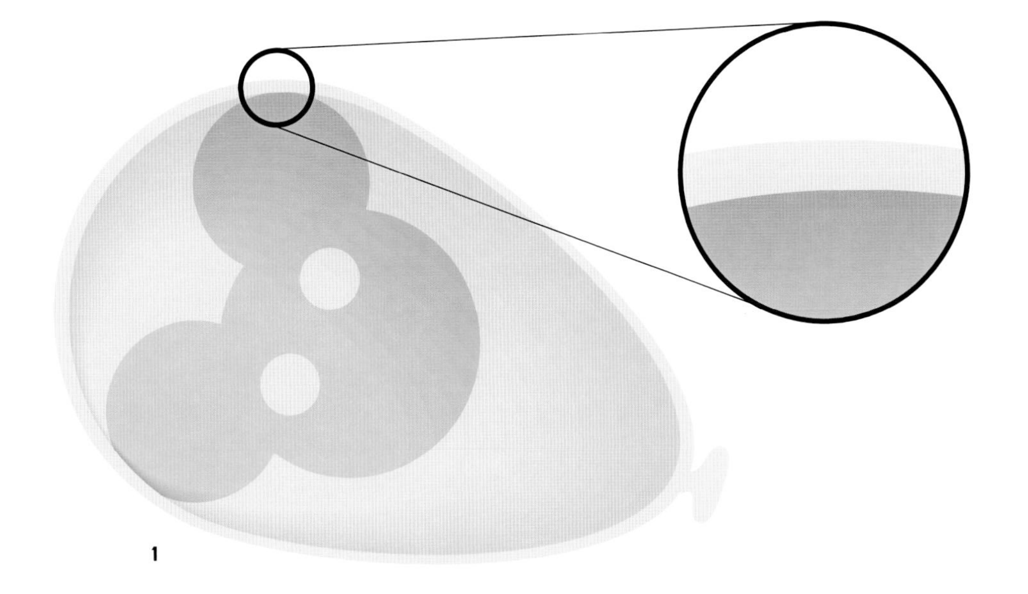

1. The more you inflate a balloon, the more its surface seems to be nearly flat.

hold of the Universe when it was a seed containing all of the mass-energy that was to become the visible Universe, but was still much smaller than an atom, and made it expand dramatically. For obvious reasons, this process became known as 'inflation'.

In everyday language, it is hard to see how there was time between time zero and 0.0001 s for anything much to happen. But remember that there is ten times as much time in 100 s as there is in 10 s. In the same way, there is ten times as much time in 0.0001 s as there is in 0.00001 s, and so on. According to simple versions of inflation theory, during the time of inflation, the seed of the Universe doubled in size once every 10^{-34} s. This means that in the span of just 10^{-32} s there would be time for 100 doublings, because, of course, 10^{-32} is 100 times bigger than 10^{-34}. Only 150 such doublings would be enough to take something that was 10^{20} times smaller than a proton and inflate it into a sphere roughly the size of a grapefruit.

It is hard to grasp what this means, because none of us has seen a proton. The best way to visualize the effect of inflation at work is to imagine the surface of a child's balloon, which is clearly bent around in a curve. Now imagine inflating the balloon to the size of the Solar System without it bursting. From the surface of the hugely inflated balloon it would be very difficult to tell that it was curved – it would seem to be flat. In exactly the same way, inflation in the first split-second after time zero could have flattened the space we live in. At the same time, the seed that inflated to become the Universe would have been too small to contain much in the way of irregularities, which explains why the Universe is so smooth.

The change brought about by inflation was very dramatic – the equivalent of inflating a tennis ball to the size of the entire visible Universe in just 10^{-32} s. Inflation predicted a Universe today that is so nearly flat we can never hope to measure any difference from flatness.

But why should such a dramatic event happen?

The Ultimate Free Lunch

The reason why people took inflation seriously is that the laws of physics provided a natural reason for it to happen. Linde and Guth (and then others) found that inflation fell out of the laws of physics without having to be looked for; it came free with the standard model of particle physics. As Guth put it, the Universe is 'the ultimate free lunch'.

Four Kinds of Force

The particle physics which gave us a free-lunch Universe along with the equations is based on an idea called 'grand unification', and the variations on the theme are known as Grand Unified Theories, or GUTs (it should be 'Grand Unified Models', but the acronym isn't so appealing). These ideas start from the fact that there are four kinds of force known in nature – electromagnetism, gravity, the 'strong force' that holds nuclei together, and another (called the 'weak force') that is responsible for radioactivity and nuclear fission.

The Holy Grail of particle physics is to find one set of equations that describes how all four of these forces behave, as different facets of some single 'superforce'. In the 1960s, physicists made the first step towards this goal by unifying electromagnetism and the weak force in one package, known as 'electroweak theory' (theory, not model, you notice, because it has been tested by experiment and passed). In the 1970s they developed a not-quite-perfect account of how to include the strong force in the package. The key thing about this work is that it showed that the forces become indistinguishable from one another at high enough energies. In everyday language, this corresponds to high enough temperatures. Looking back towards the Big Bang, when the Universe was hot enough, electromagnetism and the weak force would have merged into one force. At hotter temperatures still (by which we mean earlier times), the strong force would have merged in with them. And the guess is that at the earliest and hottest times, even gravity would have merged into the superforce.

But it is what emerges when you calculate how the forces split apart as the real Universe expanded away from time zero that made cosmologists sit up and take notice of these ideas.

Breaking the Symmetry

If the Universe was 'born' at the Planck time, 10^{-43}s after time zero, gravity would have immediately split off from the other three forces. But the strong force would not have split off until 10^{-35} s after time zero, and as it did so it would have released an enormous amount of energy. This is what gave the Universe the huge burst of expansion that we call inflation.

The breaking apart of the components of the superforce was the equivalent at this level to the kind of phase transition that occurs when water freezes into ice. A phase transition like this releases energy, which is known as 'latent heat'. This is because the rearrangement of the molecules involved takes them to a lower energy state. So even if the outside temperature is well below freezing, a bucket containing a mixture of water and ice will stay exactly at 0 degrees Celsius until all the water is frozen, even though it is losing heat to the outside world. In the same way, if you heat a mixture of water and ice, as long as there is still some ice

RIPPLES IN THE BACKGROUND RADIATION

The satellite known as COBE (left) was launched on 18 November 1989 to accurately measure the microwave background radiation from above the Earth's atmosphere. Almost immediately, the satellite measured the overall temperature of the sky more accurately than ever before, and showed that the spectrum of this radiation at different wavelengths matches the appropriate spectrum for a 'black body' (▷p. 88) so accurately that it was impossible to see any deviations from the black body spectrum.

But the main purpose of the mission was to map the entire sky, looking for tiny differences in temperature that would reveal the existence of corresponding irregularities in the material Universe at the time when the radiation last interacted with matter, between 300,000 and 500,000 years after the birth of the Universe. It took more than a year to make the observations and months to analyse the 70 million measurements made. But it was worth the wait. The analysis showed that there are hot patches on the sky just 30 millionths of a degree hotter than average, and cool spots just 30 millionths of a degree cooler than average. These 'ripples in the background radiation' correspond to the way hydrogen and helium (and dark matter) were distributed at the end of the fireball era – the seeds from which clusters of galaxies grew.

COBE showed that the background radiation is just irregular enough to account for the fact that galaxy clusters exist. Even better, the nature of the 'ripples' exactly matches the pattern predicted by inflation.

present the mixture stays at zero degrees Celsius, because the heat is simply being used to melt the ice, not to warm up the water.

In a similar way, the situation when the four forces have broken apart corresponds to a lower energy state of the entire Universe than when they formed the single superforce. It is the latent heat released when the symmetry between the components of the superforce was broken in the first split-second after time zero that drove inflation. And, like all good theories, this made a testable prediction.

The Missing Mass

Inflation says that the Universe we see around us should be so nearly flat that it is impossible for any test we can devise to measure the difference. In other words, $\Omega = 1$ precisely. This solves the puzzle of the pencil standing on its point (▷p. 109), and because the entire visible Universe must have started out, if inflation is correct, from a seed far smaller than a proton it explains why the Universe is so smooth: there was no room for anything except the tiniest irregularities in the seed.

The prediction that $\Omega = 1$ means that there must be a lot more matter in the Universe than we can see, in order to make space flat. Because it has not been seen, this is sometimes referred to as 'missing mass', although really it is the light from the mass that is missing. The alternative name for this stuff is dark matter. There must be a lot of it, because if we look at a chosen volume of space and estimate the mass of all the bright stars in all the bright galaxies, it comes out as only about 1 per cent of the critical density.

The COBE satellite should have been launched by the Space Shuttle in 1986, but was delayed for three years after the Challenger disaster.

1. The way galaxies rotate shows that they are embedded in halos of dark matter.

1. A technician working inside one of the detectors at the Large Electron Positron collider (LEP) at CERN.

2. The change from water into ice is a phase transition analagous to the change in the state of the early Universe which drove inflation.

Atoms are not Enough

You might think that a lot of what we call dark matter might in fact be faint stars and planets, or gas and dust, that we cannot see because it doesn't glow brightly. But the standard model of the Big Bang only allows for a little of this. The standard model predicts just the right amount of hydrogen and helium (75 per cent hydrogen, 25 per cent helium) – and this is what we actually do see in the oldest stars (▷ p. 18) – but only if the total density of atomic matter in the Universe is less than 5 per cent of the critical density. If the density were any higher than this, the models tell us, hydrogen and helium would not have come out of the primordial cosmic fireball in the proportions that have been observed. Even without the need for inflation, this is not enough to explain our observations of the way galaxies move in clusters, because those studies show that galaxies are in the gravitational grip of enough dark stuff to account for at least30 per cent of the critical density.

Ordinary atomic matter (the stuff of which we, the Earth, the Sun and stars are made) is called 'baryonic matter', because the protons and neutrons at the hearts of atoms are members of a family called baryons. The combination of the standard model and observations of galaxies tells us that there could be about three times (and no more than five times) as much dark baryonic matter as there is bright stuff in the Universe, and ten times as much non-baryonic matter as there is baryonic matter (both bright and dark).

Inflation says that, in order to make the Universe flat, there must be even more stuff out there. What is it?

Two Kinds of Dark Matter

Once again, particle physics comes to the rescue. The models of particle physics that describe GUTs and suggest the possibility of inflation also predict that there should be other families of particles, in addition to those detected in particle accelerator experiments. According to these models, there could be two kinds of – as yet undetected – dark matter produced in the Big Bang.

The first kind of hypothetical dark matter consists of particles which are very light (even lighter than an electron) and which were 'born' in the Big Bang. These travel at very high speeds, a sizeable fraction of the speed of light. They are referred to as 'hot dark matter', or HDM. But although it is quite likely that there is some HDM in the Universe, there cannot be enough to make it flat. The problem with HDM is that the fast-moving particles would blast apart any structure that tried to form in the early Universe, and too much HDM would stop galaxies forming at all.

The second kind of hypothetical dark matter consists of fairly heavy particles (each perhaps about as massive as a proton) which emerged much more slowly from the Big Bang. This is called 'cold dark matter', or CDM. The great advantage of CDM is that the slow-moving, massive particles would clump together as a result of their mutual gravitational attraction, forming gravitational 'potholes' into which baryonic matter would fall and accumulate, and act as the seeds from which galaxies and clusters of galaxies could grow.

There are about 10,000 cold dark matter particles in every cubic metre around us, including 'empty' space, the air that you breathe, and seemingly solid objects such as the Earth.

2

Looking for Dark Matter

There are two ways to test the prediction that the Universe is filled with a sea of dark matter particles. The first is to look for them directly. CDM particles interact with baryonic matter only through gravity or when a CDM particle collides directly with a baryon in the nucleus of an atom, making the atom jiggle about. The effect is tiny, and normally it would go unnoticed because, even in solids, atoms are constantly in motion at room temperature. But if a block of metal is cooled very close to the absolute zero of temperature (0 K), this thermal motion is greatly reduced, and the effect of a CDM impact might *just* be detected as a tiny rise in temperature of the supercold metal. According to the models, there are between one and 1000 such collisions (the exact number depends on details of the models) in every kilogram of matter (cheese, steel, water or whatever) every day. To put this in perspective: a kilogram of matter contains about 10^{27} baryons.

At the higher end of the calculated collision rate, the effect of these CDM impacts could just be detected using modern technology. Experiments searching for dark matter in this way are now underway, buried deep down mine shafts to shield them from interference. Because they are exploring the limits of what is technically possible, nobody will be surprised if they do not produce a positive detection. But if they do, direct evidence for the dark matter that fills the Universe will have been found at the bottom of a hole in the ground.

The second way to test the prediction takes us back out into the exploration of the Universe at large. Computer models are now well able to describe how ordinary baryonic matter clumps together to form galaxies in a variety of different cosmological models, with different mixtures of dark matter in them. These simulations show that if there were only baryons in the Universe, and if the density was just 5 per cent of the critical density (that is, $\Omega = 0.05$), then there is no way that the pattern

IN THE GRIP OF THE DARK STUFF

The speeds at which different parts of a galaxy rotate can be measured by the Doppler effect, which shows that galaxies like the Milky Way rotate differently from the way planets orbit the Sun. In the Solar System, the outer planets not only take longer to complete a single orbit, but actually move more slowly through space than the inner planets, with orbital speed decreasing as the square root of the distance of a planet from the Sun. In a galaxy, the stars sweep around at the same speed, no matter how far they are from the centre of the galaxy. Stars that are further out still take longer to complete one orbit, because they have further to travel. But they all travel at the same speed.

The only way this can be explained is if all of the visible stuff in a disc galaxy is embedded in some larger dark halo, and is held in the gravitational grip of the dark stuff in the halo. The evidence shows that there must be about ten times as much of this dark stuff as there is matter in the form of bright stars in a galaxy like our own. The way whole galaxies move about in clusters shows that there is even more dark matter present in the seemingly empty spaces between them.

1. (opposite) The face-on spiral galaxy NGC1232.

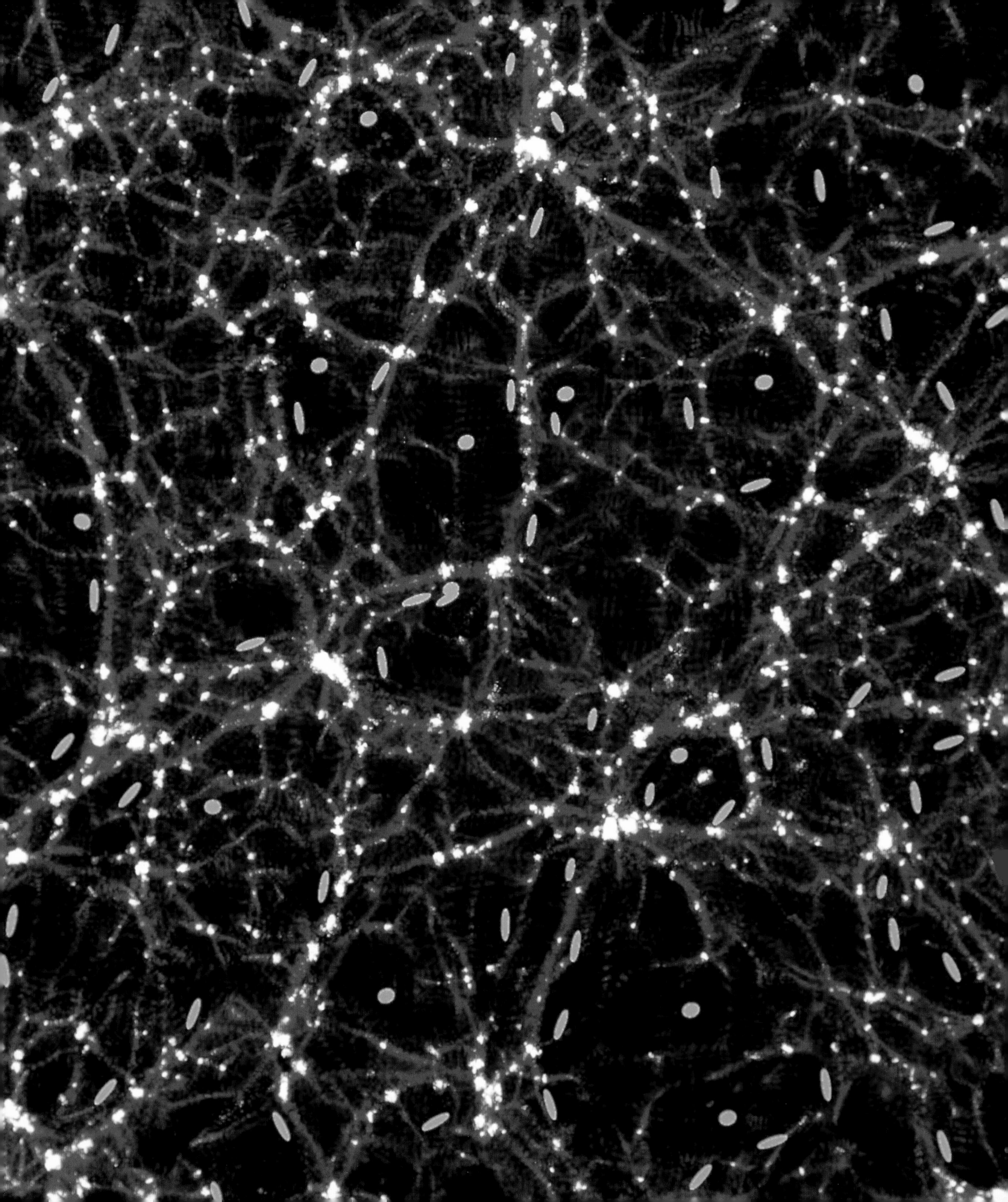

of galaxies and clusters of galaxies that we see around us could have formed in the time available since the Big Bang. Adding enough HDM to flatten the Universe makes things even worse. But if there is at least 30 per cent of the critical density in the form of CDM, then the pattern of galaxies and clusters that emerges from the simulations is strikingly similar to the pattern we see in the real Universe. This is powerful evidence that CDM is the dominant form of matter in the Universe, although these results do allow for the possibility that there is a little HDM as well (perhaps more HDM than there is baryonic matter).

This picture of a Universe dominated materially by CDM was firmly established by the end of the 1990s. But it still isn't quite the last word. In order to make galaxies and clusters that look like those in the real Universe, you need less than half of the critical density in the form of dark matter – but the models still work best if that matter is embedded in flat space. If the density of matter is less than half the critical density, how can space be flat? The answer takes us back to Albert Einstein's original thoughts on cosmology, but also right up to date with key results from the astronomical frontier announced at the beginning of the twenty-first century, seeming to slot the last piece into the cosmological jigsaw puzzle.

The surprising implication of this work, yet to be confirmed, is that the Universe may be precisely flat, but it is also expanding faster as time goes by.

1. (opposite) Computer model of dark matter (red) distribution, the first time the invisible dark matter has been mapped.

2. Experiments designed to detect dark matter particles have to be buried deep underground to minimise interference.

2

WHY THE SKY IS DARK AT NIGHT

You can make one of the most profound observations in the whole of science by going outside and looking at the dark night sky. Why is the sky dark at night? Is it because there are gaps between the stars? But if space went on to infinity, and it was filled with stars, every 'line of sight' out into space would end on a star, and the whole sky would blaze with light. So the Universe cannot be infinite and uniform.

1

Olbers and the Edge

The puzzle of the dark night sky often goes by the name 'Olbers' paradox', after the nineteenth-century German astronomer who publicised it. But he wasn't the first to ponder the puzzle, and it isn't really a paradox. The simplest way to picture the puzzle is to imagine standing in a large forest. Wherever you look, you will see a tree. But if you are only standing in a small wood, you might be able to see through the spaces between the trees, out to the edge of the forest. In an infinite Universe, everywhere you look you will see a star. The fact that we can see 'out' through the gaps between the stars seemed to imply, to Olbers and others, that there must be an 'edge' to the Universe, beyond which there was only dark, empty space, and no more stars. This reasoning applies even if you think in terms of galaxies rather than stars.

Looking Back in Time

The first person who realized that this need not be the case was Edgar Allan Poe who, alongside his literary activities, was a keen amateur scientist. He gave a lecture setting out the correct resolution to Olbers' paradox in February 1848. But he died a year later and no scientists took up his suggestion.

What Poe pointed out was that by looking further out into space we are looking further back in time, because it takes light a finite time to travel through space. When we look through the gaps between the stars at the dark night sky, we are looking back in time to an era before the stars were born – in Poe's words, to a distance 'so immense that no ray from it has yet been able to reach us at all.'

The Edge of Time

This notion needs only a tiny modification to fit in with the idea of the Big Bang. The point is that the Universe is not infinitely old, and it has an 'edge' in time (the Big Bang) rather than an edge in space. Looking through the gaps between the stars and galaxies, we do indeed look back to a time before galaxies formed.

But modern instruments reveal that the sky is not completely dark. What we 'see' there is the background radiation from the Big Bang fireball itself, once as hot as the surface of a star, but now redshifted by the expansion of the Universe down to 2.7 K.

The darkness of the night sky is evidence that the Universe was born at a definite moment in time. You can see (or rather, not see!) evidence to support the Big Bang model with your own unaided eyes.

1. In an infinite universe, you ought to see stars everywhere you look, just as in an infinite forest you would see trees everywhere you look.

2. Edgar Allan Poe was the first person to appreciate why the sky is dark at night.

TOPIC LINKS

2.1 The Big Bang
p. 83 Microwaves from the Birth of Time
p. 88 Black Body Radiation

2.2 Cosmology for Beginners
pp. 102–3 The Arrow of Time

2.4 THE ACCELERATING UNIVERSE

One of the cherished features of the standard Big Bang model that had developed in the 1970s and 1980s was that the gravitational pull of all the matter in the Universe must be making the universal expansion slow down as time passes. Yet, in 1998 the discovery that the Universe is expanding *faster* as it ages earned the accolade 'breakthrough of the year' from the journal *Science*, and was presented to the public as a stunning achievement which overturned conventional cosmological wisdom. But the hype did less than justice to previous generations of cosmologists. The possibility of universal acceleration was already built into the cosmological equations developed by Albert Einstein back in 1916. In fact, to the cosmologists themselves it turned out to be just what the doctor ordered – the last piece in the cosmological jigsaw puzzle. But confirmation had to wait for another couple of years.

EINSTEIN'S UNWELCOME CONSTANT

Albert Einstein tried to describe the Universe mathematically, using his general theory of relativity, in 1917. He wanted to represent the simplest possible model, in which matter is uniformly distributed through space. He also wanted this model to be static, neither expanding nor contracting, to match the fact that the Milky Way (which was thought to be the entire Universe at that time) is neither expanding nor contracting. The only way in which he could achieve this was to include in his equations the term now known as the 'cosmological constant', and denoted by the Greek letter lambda (Λ). The equations said nothing about the value of this constant – they allowed for the possibility that it is zero, or it could have any positive or negative value. Depending on the size of the constant, it could act like antigravity, holding matter up against the inward pull of gravity, or like an addition to gravity, pulling things together even more effectively. Einstein chose a value that held his model still, in a sense cancelling out gravity, and in the very last sentence of his first paper on cosmology, published in 1917, he wrote, 'That term is necessary only for the purpose of making possible a quasi-static distribution of matter, as required by the small velocities of the stars.'

When Hubble and Humason discovered that the Universe was expanding, Einstein said that the introduction of the cosmological constant had been the biggest blunder of his career. But other researchers took the term more seriously.

1. Albert Einstein's general theory of relativity is the basis of our understanding of the Universe.

Exploring the Models

Einstein had been looking for a single solution to the equations of the general theory of relativity, a unique model which corresponded to the real Universe. But the equations actually offer a variety of models. The first person to appreciate this was the Russian Alexander Friedmann, who was also the first person to incorporate expansion as an integral feature of the cosmological models that he explored mathematically.

Friedmann published his solutions to the cosmological equations of the general theory

1. Edwin Hubble at the controls of the Hooker telescope.

in 1922. He understood that there is no unique solution to those equations (as Einstein had hoped) but rather a family of models describing different ways in which spacetime could evolve – different model universes. There was no way at the time to tell which, if any, of the models corresponded to our Universe. Significantly, however, all of Friedmann's models incorporated expansion at some phase in their evolution.

In some variations on the theme, space expanded forever. In others, it expanded for a time and then re-collapsed. Sometimes the expansion was fast and sometimes it was slow. It is even possible to find models which start out very big, shrink down to a certain density and then expand. But in all of the models there was at least some phase of the evolution in which the model universe expanded in such a way that from any point in it you would see other points receding with velocities proportional to their distances – exactly what Hubble and Humason found before the end of the 1920s.

Models for All Tastes

Einstein was never really happy with this variety of models, and continued to seek a unique, special case that might describe the real Universe. Together with the Dutch astronomer Willem de Sitter, in the early 1930s (soon after the discovery of Hubble's Law), he developed the Einstein–de Sitter model, which is the simplest variation on the theme allowed by the equations of the general theory. It is the model in which space is precisely flat (the only special case Einstein could think of) and with (= 0. This became the benchmark against which other models were compared.

But there was one embarrassing feature of the Einstein–de Sitter model, that Einstein and de Sitter were careful not to mention. In that model, there is a unique relationship between the rate at which the Universe is expanding today and its age – obviously, the faster it is expanding now, the less time it has taken to reach its present size , but you also have to allow for the way in which the expansion has slowed down since the Big Bang. Using the value of the constant in Hubble's Law (the redshift-distance relationship) found by Hubble himself, this gave the age of the Universe (assuming it really was described accurately by the Einstein–de Sitter model) as just 1.2 billion years, only about a third of the age of the Earth, which was already very well known by the 1930s.

Clearly, something was wrong. We now know that the early measurements of Hubble's Constant were much too big and that the true age of the Universe is about 14 billion years. But in the 1930s (and the decades that followed), there was another way to resolve the dilemma, one favoured by Georges Lemaître. With a suitable choice of the equations described a model which started out from a very dense state and expanded for a while, then 'hovered', neither

The amount of computer memory needed to analyse the data from a Boomerang flight is 240 Gigabytes. The amount needed to analyse the Planck data will be 1600 Terabytes.

THE FATE OF THE UNIVERSE

The expansion of the Universe is getting faster, because the springiness of space (▷ p. 131) is acting to overwhelm the inward tug of gravity. The effect gets bigger the more space there is, so long ago it had very little influence on the way the Universe expanded. That is why it is not obvious from redshift studies that there is a cosmological constant. But as the galaxies get further apart (like these very distant galaxies, right), it is harder for gravity to pull them back together, and as the I effect increases it will eventually dominate, making space expand faster and faster and taking galaxies along with it. Even though the Universe is flat, the very 'dark energy' of the vacuum that makes it flat also ensures that it will expand forever, at an increasingly rapid rate (this could happen even if space were closed, like the inside of a black hole).

Although the amount of matter in the Universe stays the same, as a proportion of the critical density at each epoch (30 per cent today) it decreases, while the proportion of dark energy increases at exactly the right rate to compensate. In the past, gravity dominated. In the future, the I will dominate. We live at an unusual moment in the history of the Universe, when the dark energy and the mass energy components are roughly the same size – the mass in dark energy is only twice that in matter. Nobody knows why this should be so, and it may just be a coincidence. It is certainly hard to see how life forms like us could survive in the far future of the Universe, when all the stars have burnt out and all that remains are the ashes, in the form of white dwarfs, neutron stars and black holes, forming black galaxies being carried ever further apart by expanding black space.

1. The two giant Keck telescopes on Mannakea in Hawaii. Each telescope has a mirror ten metres across.

2. Werner Heisenberg, whose uncertainty principle may hold the key to understanding the birth of the Universe.

1

expanding nor contracting, for an indefinite interval before beginning to expand again. If our Universe were like that, and we lived in the second phase of expansion, it might be much older than implied by measuring the redshift-distance relationship today. Which kind of model you chose to describe the Universe seemed, in the 1930s, just a matter of personal preference.

Taking Λ Seriously

This is just one example of how choosing a value of the cosmological constant could resolve any problem that cropped up in cosmology, such as the age problem. Mathematicians enjoyed exploring the possibilities, but astronomers tried to do without the term, not least because it seemed too easy to use it as a fudge factor that could be adjusted to match any requirement of the observations. But when the observations of the real Universe began to be good enough to eliminate many of the wilder flights of cosmological fancy, it became clear that even with the up-to-date estimates of the Hubble Constant and the correspondingly extended age of the Universe, something was missing from the Einstein–de Sitter model. The cosmological constant came in from the cold in the 1990s, not so much because astronomers wanted it, but because they had no choice. They had, following the dictum that Conan Doyle put in the mouth of Sherlock Holmes, 'eliminated the impossible', and what they were left with, no matter how improbable, had to be the truth.

Speeding Supernovae

By the late 1990s, the idea of inflation (▷ pp. 114–15) was looking very good, especially in the light of the COBE data and other measurements of the microwave background. The Universe must be flat. But at the same time, studies of the way galaxies move had been unable to come up with firm evidence for more than 30 per cent of the matter needed to make space flat (▷ p. 95). One way out of the dilemma, which began to be taken increasingly seriously in the second half of the 1990s, was the Λ term.

Putting a Spring in Space

The cosmological constant actually has two effects on the Universe. First, with the right value of Λ, it makes space springy, producing an antigravity effect, a kind of cosmic repulsion. This is equivalent to an energy of empty space, similar to the way the force of gravity is associated with the energy of matter.

This is where the second effect of Λ on the Universe comes in. Because energy is equivalent to mass, from $E = mc^2$, the Λ term also has a gravitational influence. With the right choice of Λ, it is possible to have a universal energy (associated with the cosmic repulsion) that amounts to 70 per cent of the mass needed to make the Universe flat, and at the same time to have only a tiny effect on the expansion of the Universe, one that is barely detectable today.

Adding the 70 per cent from the Λ term to the 30 per cent in the form of matter gives just the right amount of mass-energy to flatten the Universe. The theorists had explored the possibilities allowed by the models and found the simplest way to make everything fit together if there really is only 30 per cent of the critical density around in the form of dark matter and inflation did occur. It wasn't as simple as many people had hoped and it still looked a bit contrived but, like all good theories, it could be tested by new observations of the real Universe.

The Supernova Story

The role of supernovae in the story is as standard candles that provide a way to measure distances far across the Universe (▷ p. 73). Type I supernovae don't all have precisely the same brightness, but their absolute peak brightness can be inferred from the way they fade away after reaching a maximum. We can see such supernovae in nearby galaxies whose distances are known, so when they are detected in far-distant galaxies their distances can be worked out by comparing their apparent brightnesses with the nearby supernovae. The difficulty is finding supernovae in very distant galaxies, and it was only in 1998 that the technology was good enough to do the job. Then, two teams exploited that technology to make independent studies of the same phenomenon. Happily, they both got the same answer.

Two international teams, one using the Keck telescope in Hawaii and the other based at Australia's Mount Stromlo and Siding Spring Observatories, measured the brightnesses of dozens of Type I supernovae in very remote galaxies and compared the distances they inferred with the redshifts of those galaxies. They found that the recession velocities of those galaxies are a little less than would be expected by applying Hubble's Law, calculated from nearby galaxies, to them.

This means that the expansion of the Universe is accelerating, not decelerating. The point is that we see *nearby* galaxies as they were quite recently, but we see *distant* galaxies as they were long ago, because the light from them has taken so long to reach us. Nearby galaxies are receding from one another faster than distant galaxies are, because the expansion of the Universe is getting faster.

On its own, this discovery would have been interesting enough. What made it spectacular is that the amount of cosmic repulsion required to match the observations is also exactly right to provide 70 per cent of the mass-energy needed to flatten the Universe.

Clinching Confirmation

The one remaining uncertainty was whether the Universe really is precisely flat,

Searches for distant supernovae have to be carefully timed because these objects are so faint that they can only be observed properly at New Moon, when the sky is darkest.

2

as inflation says it should be. Again, new technology provided a way to test the prediction. Radiation travelling through space is affected by the curvature of space, and the further it travels the more it is affected. The microwave background radiation has been travelling through space for longer than anything else we can detect, and the exact pattern of the variations in that radiation from one part of the sky to another can, in principle, reveal the curvature of space all the way from the Big Bang fireball to us – across 14 billion light years of space.

By the end of the 1990s, instruments carried aloft in balloons were able to monitor fluctuations in the background radiation 35 times smaller than those COBE could detect. Results from two of those balloon flights, announced in 2000, show that the Universe is flat to within 10 per cent – which means that lies between 0.9 and 1.1. With clear evidence that there is only 30 per cent of the mass needed to flatten the Universe around in the form of matter, that meant there had to be 70 per cent around in the form of energy – just as the supernovae suggested. It was compelling evidence that the cosmological constant is real, whether Einstein liked it or not.

Ripples in the Cosmic Sea

The close agreement between all these different elements of the story – the amount of matter in the Universe, the acceleration required by supernova studies, the flatness revealed by measurements of the background radiation – makes inflation far and away the cosmological 'best buy' at the beginning of the twenty-first century. This provides a way to tackle the last great puzzle: why the Universe is not perfectly uniform, but has irregularities big enough to allow us to exist.

Quantum Fluctuations

Where does the energy of empty space come from? According to quantum physics, there is no such thing as truly 'empty' space, because that would have precisely zero energy, and one of the most famous rules of the quantum world, Heisenberg's Uncertainty Principle, says that it is impossible for anything to have a precise value (not just that it is impossible for us to measure things precisely; absolute precision does not exist in the Universe). In this context, in any tiny volume of space there is a tradeoff between time and energy. Particles known as 'virtual pairs' can (indeed, *must*) pop into

INTO THE FUTURE

The definitive answers to most of the remaining questions in cosmology should come when the instrument technology used in projects such as Boomerang is harnessed to space technology. A new generation of microwave satellites needs to be put into orbit around the Earth. There they will be able to map the whole sky at microwave frequencies, staying in orbit for months or years to pick out the fine details.

The first of these satellites, the Microwave Anisotropy Probe (or MAP, right) was due to be launched by NASA shortly. If successful, it will spend a year mapping the sky. Then the team will spend a year analysing the data. MAP is a quick and relatively cheap mission which will have about the accuracy of Boomerang but will extend to cover the whole sky.

The latest word in microwave mapping yet planned is the European Planck mission (named after Max Planck), which should fly into space before the end of the first decade of the twenty-first century. It is a more ambitious project which, if all goes well, will provide more accurate data than anything that has gone before. Together, these missions should pin down the values of the key cosmological parameters, including Λ, Ω, and Hubble's Constant, to an accuracy of 0. 1 per cent.

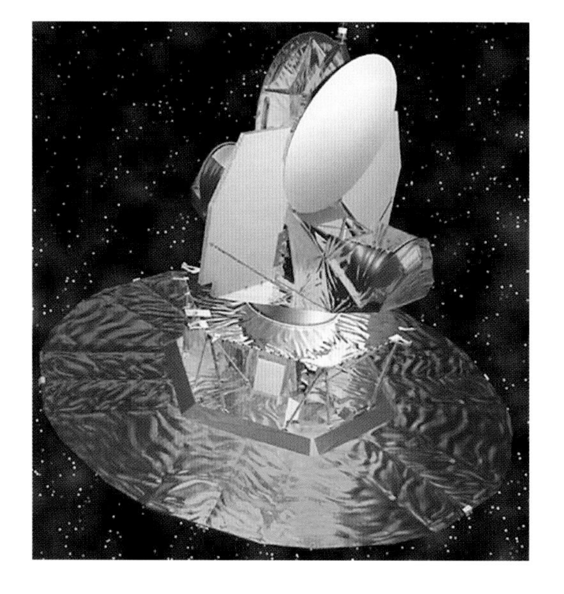

existence out of nothing at all, provided they annihilate one another and disappear within a certain time. The time limit is set by their mass – the bigger the mass, the smaller the time they can exist for – but is measured in tiny fractions of a second. It is as if the particles exist when the Universe isn't looking, but as soon as it has time to notice their presence, they vanish.

The effect of all this is to make space a seething foam of virtual particles, giving it both energy and structure. It is the energy that provides the outward push of the term, and the mass associated with that energy that completes the job of making the Universe flat.

It is no coincidence, by the way, that all the different kinds of mass-energy in the Universe add up to make it flat, with $\Omega = 1$. Inflation drives the Universe to flatness, so there is only so much mass-energy to go round, to be shared among baryons, HDM, CDM and Λ. It's as if you had a variety of different bottles and jars and you poured water into them from a one-litre container. No matter how the water is divided among the different containers, the total amount of liquid has to add up to one litre.

But although quantum fluctuations are usually ephemeral, it seems they have left their imprint on the Universe.

1. In a fractal, each small piece of the pattern can be enlarged to reproduce the whole pattern.

A Matter of Scale

As well as occurring in tiny split-seconds of time, quantum fluctuations also occur on tiny distance scales (not least because the disturbances involved do not have time to travel far before they are forced to disappear). These processes were going on in the earliest phase of the Universe, after the Planck time and before inflation. At the time inflation took a grip on what was to become the visible Universe, all of the mass-energy associated with the visible Universe was tucked inside a tiny seed just 10^{-25} cm across – 100 million times bigger than the Planck length but still 1000 billion times smaller than a proton. Even this ridiculously tiny seed was big enough to contain quantum fluctuations, involving energetic fields (like electromagnetism) rather than particles. So the vacuum had an ever-changing structure, but the structure always matched a certain statistical pattern.

Then inflation happened. Everything in the universal seed was ripped apart and spread out. In the process, whatever fluctuations of the vacuum were going on at the moment inflation began were frozen into the structure of the rapidly expanding seed and also enormously stretched out as space expanded. Space actually expanded faster than light during inflation (this is entirely allowed by Einstein's equations; it is only motion *through* space that cannot exceed the speed of light), and the last quantum fluctuations were imprinted on the pattern of hot gas that emerged from the cosmic fireball.

The statistical pattern of the quantum fluctuations is called 'scale invariance', because it looks the same (statistically) on all scales – if you take a piece of the picture and expand it, it doesn't look precisely like the original, but has the same statistical appearance, in terms of the arrangement of hot spots and cold spots. What COBE and its successors see in the ripples in the background radiation is exactly the same pattern of scale invariance, but 'written' over hundreds of millions of light years, instead of within a sphere 1000 billion times smaller than a proton. We are part of that pattern – life is part of the structure imprinted on the Universe by quantum fluctuations shortly after the birth of time.

BALLOONS AND THE BACKGROUND RADIATION

Technology has improved so much since the time of COBE that, in the absence of any new microwave satellites, the best maps of the microwave background radiation now come from instruments attached to balloons on flights above most of the Earth's atmosphere. Until the next generation of microwave satellites goes up, taking comparable instruments with them right into orbit, the best picture of the microwave Universe comes from one of those experiments, Boomerang.

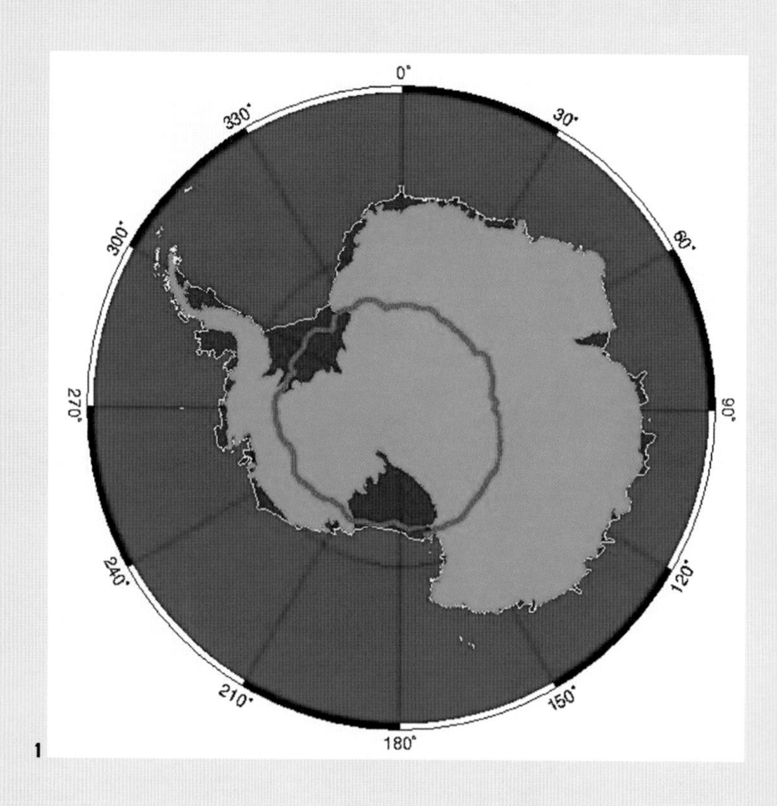

1

2

Around the World in 10.5 Days

Boomerang gets its name from the fact that the balloon travels in a roughly circular path around Antarctica, carried by high altitude winds. Technically, because it circles the South Pole, this is a round-the-world flight. Its first circumpolar flight started from the McMurdo base at 03:30 GMT on 29 December 1998, and ended when the balloon returned to its starting point at 15:50 GMT on 8 January 1999. It went around the world in 10.5 days.

The microwave telescope used to monitor the background radiation was lifted to an altitude of 40 km by a balloon the size of an American football stadium.

Why Antarctica?

Antarctica is a very good place to operate a balloon mission like Boomerang for several reasons. First, because the path of the balloon is predictable and it returns to its starting point, it can stay aloft for a long time. It cannot stay up as long as a satellite, but similar balloons launched in other parts of the world have to be brought down in a few hours (at most a couple of days) before they become a hazard or get lost.

Secondly, the air above Antarctica is cold and dry, which means that even the remaining trace of atmosphere above 40 km altitude does not have much influence on the incoming background radiation, which can be partly absorbed by water vapour.

Finally, because Antarctica is uninhabited (except by scientists and penguins) the balloon is not going to get in anyone's way, and there are no local radio and television stations to interfere with its microwave detectors.

Supercool Science

Even with all these advantages, mapping the

microwave sky accurately is difficult. Basically, the detectors are measuring the temperature of the background radiation from different parts of the sky, and they can only do this accurately if the detectors are even colder than the radiation (the colder the better). The radiation has a temperature of 2.735 K. The Boomerang detectors were cooled to 0.28 K (–272.88 K2, since 0 K is –273.16 degrees Celsius) in a giant Dewar (like a Thermos flask) at the focus of a 1.3 metre diameter telescope.

The Boomerang Dewar contained 65 litres of liquid helium in an inner container and 75 litres of liquid nitrogen in an outer container. Together these are sufficient to keep the detectors at the required temperature for up to 12 days.

Confirmation by Balloon

Although other launch sites are less ideal than Antarctica, other balloon experiments provided valuable confirmation that Boomerang really is detecting fluctuations in the background radiation, and not some spurious 'signals' caused by problems with its detectors. The most important of these complementary balloon programmes was dubbed MAXIMA, from Millimeter Anisotropy eXperiment Imaging Array.

The first flight of MAXIMA took place in August 1998, and lasted for just 4.5 hours. It uses a similar 1.3 metre telescope to the one on Boomerang, cooled to about 0.1 K. On this and subsequent flights, MAXIMA found the same kind of fluctuations as Boomerang, but in the northern sky (it was launched in Texas) not the southern sky. Although the results were not as impressive in themselves as those from Boomerang, they were crucially important because they showed the same pattern from different parts of the sky, confirming that what Boomerang saw is a universal effect.

3

1. Map of the flightpath of the Boomerang.

2. Boomerang's map of the microwave sky. The colour variarions represent varaitions in temperature.

3. The launch of MAXIMA in August 1998.

TOPIC LINKS

2.1 The Big Bang
p. 83 Microwaves from the Birth of Time
p. 84 The Very First Light
pp. 90–1 The First Four Minutes

2.3 Missing Mass & the Birth of Time
p. 116 Ripples in the Background Radiation

3

MAKING CONTACT

3.1 LIFE AND THE UNIVERSE

To an astronomer, exploring space (even by proxy) can be a satisfying life's work. Understanding the nature of the stars and galaxies on the basis of the laws of physics determined here on Earth, plus the information gleaned from telescopes, is a breathtaking achievement and a rewarding experience. But at the back of the mind of even the most dedicated astronomer lurks the big question – are there other astronomers out there, looking at the same stars and galaxies from a different perspective, and drawing their own conclusions about the nature of the Universe? Is there life – and in particular, is there intelligent life – beyond the Solar System? After centuries of speculation, it is a question that now seems ripe for the answering. Travelling to the stars may still be a pipe dream; but, for the first time, we have the technology to detect other worlds like the Earth, and perhaps to communicate with them.

Previous page. Ariel view of the Arecibo radio telescope in Puerto Rico.

OTHER WORLDS

As long as we only knew of the existence of one family of planets – the one orbiting around our own Sun – the possibility existed that the Solar System was unique, and that no other planets existed anywhere in the Universe. It seemed unlikely, but it *might* be true. That changed in 1995, when a team of Swiss astronomers discovered a giant planet, not unlike Jupiter, but with about half as much mass, orbiting around a star known as 51 Pegasi. It lies 50 light years from Earth in the direction of the constellation Pegasus. They didn't see the planet directly, but inferred its presence from the way the orbiting planet made the star wobble to and fro slightly, as it was tugged by the planet, moving in its orbit, first in one direction then the other. To measure the wobble of the star they used the two most valuable tools the astronomer has for the exploration of the Universe: spectroscopy and the Doppler effect (▷ pp. 18 and 82).

It's no accident that the discovery was made in the mid-1990s, and not before, because that was when spectroscopic techniques became accurate enough, and the computers used to analyse the spectra became powerful enough, to measure the tiny Doppler shifts involved. What happens is that, when a planet is on 'our' side of a star, it tugs the star towards us, producing a tiny

1. Artist's impression of the giant planet 51 Pegasi B orbiting its parent star 51 Pegasi.

blueshift in the light from the star. When the planet is on the other side of the star, it tugs the star away from us, producing a tiny redshift (▷ p. 82). In round terms, the change in speed of the star that has to be measured in this way is about 10 metres per second – almost exactly the speed of an Olympic-standard 100-metre sprinter. Measuring the Doppler shift in the light from 51 Pegasi is like measuring the speed of the winner of the Olympic 100 metres from a distance of 15 parsecs.

The Big Surprise

The star 51 Pegasi had been chosen for investigation using the Doppler technique because it is a yellow star rather like the Sun. The most important feature of the discovery was that it showed that other Sun-like stars do have planets. In that sense, the Solar System is not unique. But there was a big surprise. Like all explorers entering new territory, the planet hunters had found the unexpected. Although the planet orbiting 51 Pegasi is a giant like Jupiter, it orbits much closer to its star than any giant in our Solar System orbits the Sun (closer even than Mercury), and the heat of its star must raise the surface temperature to 1300 K. Whereas Jupiter takes just over 11 years to orbit around the Sun once, the newly discovered planet takes just over four *days* to orbit its star once.

This was such a surprise that many other astronomers wondered if the Swiss team had made a mistake. But soon their observations were confirmed by a team of researchers in the United States, and then by other observers. Within the next two years, four more 'hot Jupiters' were found orbiting other stars, as well as four Jupiter-like planets in orbits that seemed more normal, from the perspective of our Solar System. But the questions remained. How could *Jupiter*-like planets get into orbits so close to their stars? And which kind of planetary system really is normal – the Solar System, or systems like 51 Pegasi?

A Profusion of Planets

The best answer to the first question seems to be that giant planets may form far away from their parent star (as in the Solar System) but get dragged in towards it if there is still enough dust around in the disc in which they form (▷ p. 39) to provide enough friction to make them fall inwards. If there is less dust,

THE FIRST 'VISIBLE' PLANET

In the summer of 1999, astronomers from St Andrews University in Scotland and Harvard in the United States announced that they had 'seen' a planet beyond our Solar System for the first time. The planet orbits the star Tau Bootis, chosen for investigation because it is similar in size to our Sun, but rather brighter at certain wavelengths of light. The planet now thought to be orbiting Tau Bootis is too far away to be photographed in any way, but light from the parent star is reflected from the planet, and is detected, mixed up with the raw starlight, by telescopes on Earth. The reflected light from the planet is only one ten-thousandth as bright as the direct light from the star, but it can be picked out from the starlight because the planet is orbiting around Tau Bootis at 150 km per second. This produces a rhythmically changing Doppler shift in the light reflected from the planet.

This first directly observed extra-solar planet has about 1.4 times the mass of Jupiter, the largest planet in our Solar System, and is almost certainly, like Jupiter, a gas giant. If confirmed, this will be one of the most important astronomical discoveries; but confirmation of these claims has not yet come in.

and less friction, they stay where they are.

The honest answer to the second question is 'We don't know'. But there is one point worth bearing in mind. It is easier to detect big planets than little planets, and easier still to detect them if they are in close orbits around their parent stars (the closer they are, the bigger the wobble they cause). So systems like 51 Pegasi are the easiest kind of systems to find and it should be no surprise that astronomers found them first. Systems like our Solar System are harder to detect; they may be there, but our instruments are not yet good enough to spot them using the Doppler technique. As astronomers are fond of putting it, 'absence of evidence is not evidence of absence'. But so many 'extra-solar' giant planets are now known that it is hard to believe there are not a few small planets out there as well. By the end of the 1990s, 28 of these giants had been discovered, three of them in orbit around the same star, Upsilon Andromedae. To get some idea of the chances of Earth-like planets being out there as well, we have to take stock of what it is that makes the Earth special in our own Solar System.

1. Artist's impression of planet orbiting the star Tau Bootis. A hypothetical moon has been included in this rendition.

1

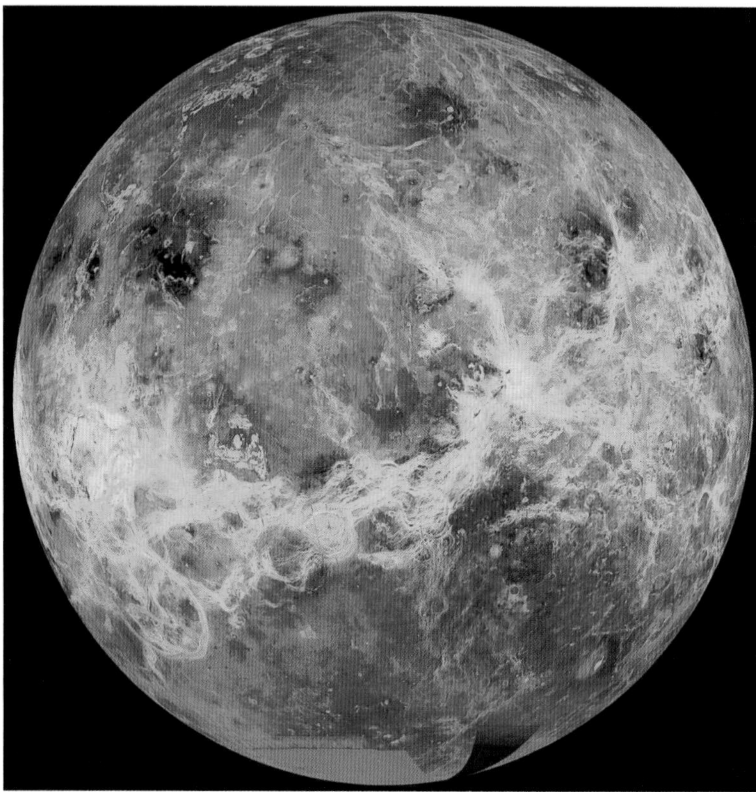

2

THE GOLDILOCKS PLANET

Earth is sometimes referred to as the 'Goldilocks' planet, because conditions here are just right for life, just as baby bear's porridge was just right for Goldilocks. The best way to see this is in terms of water (an essential ingredient for life as we know it) and temperature.

In those terms, the Earth lies roughly in the middle of the 'comfort zone' around our Sun. The planet Venus is almost a twin of the Earth in size, but its distance from the Sun is a bit less than three-quarters of the distance of the Earth from the Sun. If Venus had started out a little further from the Sun, it might have formed oceans of water in which life could emerge. But the extra heat present where Venus formed meant that liquid water could not exist there, but built up in the atmosphere as water vapour, helping to contribute to an intense greenhouse effect which, together with the Sun's proximity, sends temperatures soaring above 700 K.

Mars is a lot smaller than the Earth, with only a tenth of the Earth's mass. It orbits one and a half times as far away from the Sun. If the Earth were in the orbit of Mars, it might still be a fertile planet, thanks to the greenhouse effect of its atmosphere. But the gravity of Mars was too feeble to hold onto a thick atmosphere, and the combination of a thin atmospheric blanket and its distance from the Sun sends its night-time winter temperatures plunging below 162 K (minus 111 degrees Celsius).

Stretching things to the limit, you could say that the comfort zone around the Sun extended from the orbit of Venus to the orbit of Mars, with the Earth roughly in the middle of the zone. If this were right, life might be rare. There would be very little

1. The Earth is just right for life, whereas Venus (2) is too hot, and Mars (opposite) is too cold.

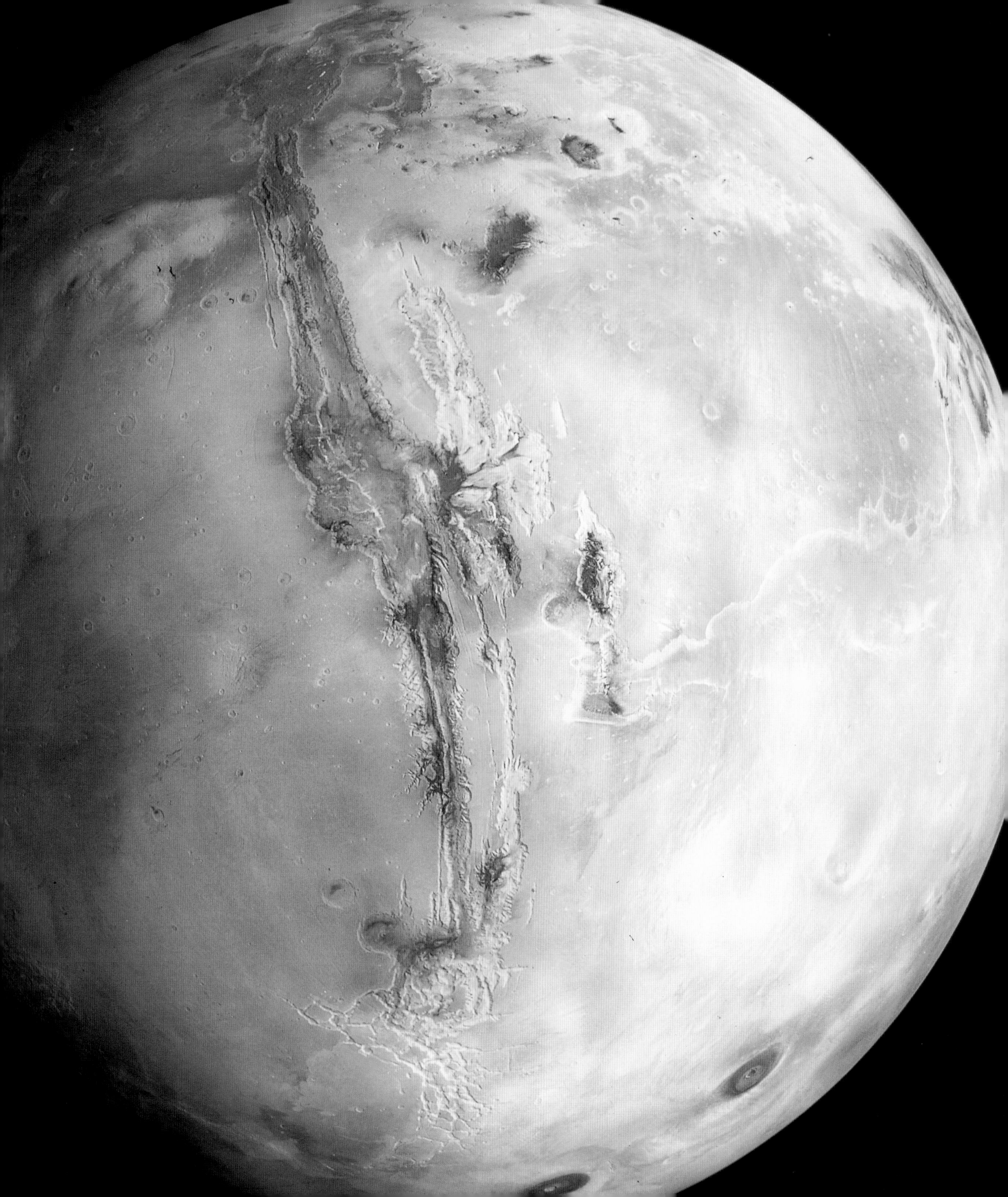

chance of finding more than one planet in the equivalent comfort zone around another star, and all kinds of things might go wrong to prevent life getting a grip on that one suitable planet. But when spaceprobes visited the outer part of our own Solar System, they found that this pessimistic point of view might be wrong.

Extending the Comfort Zone

Assuming that life elsewhere resembles life on Earth (which is the only basis we have to go on), the search for life is essentially the search for liquid water. Without liquid water, there is no life. It is no coincidence that our word for a region on Earth devoid of life, desert, is the same as our word for a region devoid of liquid water.

At first sight, it looks as if the surface of any planet beyond the orbit of Mars should be too cold for liquid water to exist. But in the late 1990s, the space probe Galileo sent back pictures from one of Jupiter's moons, Europa, which revealed that the moon is almost entirely covered by what seems to be a layer of ice floating on an ocean of liquid water, very like pack ice floats on the Arctic Ocean. Although it is smaller than our own Moon, Europa still has a diameter of 3,138 km, making it a respectably sized potential home for life. But what keeps it warm?

The answer seems to be that as Europa follows its orbit around Jupiter, the gravitational pull from Jupiter itself, and from the other moons orbiting around the planet, combine to produce a changing tidal force, constantly 'kneading' the inside of Europa. It is squeezed rhythmically in and out, in the same way that the gravitational pulls of the Sun and Moon cause the tides to move rhythmically up and down the seashore on Earth. This kneading generates heat, enough heat to melt the ice which makes up the bulk

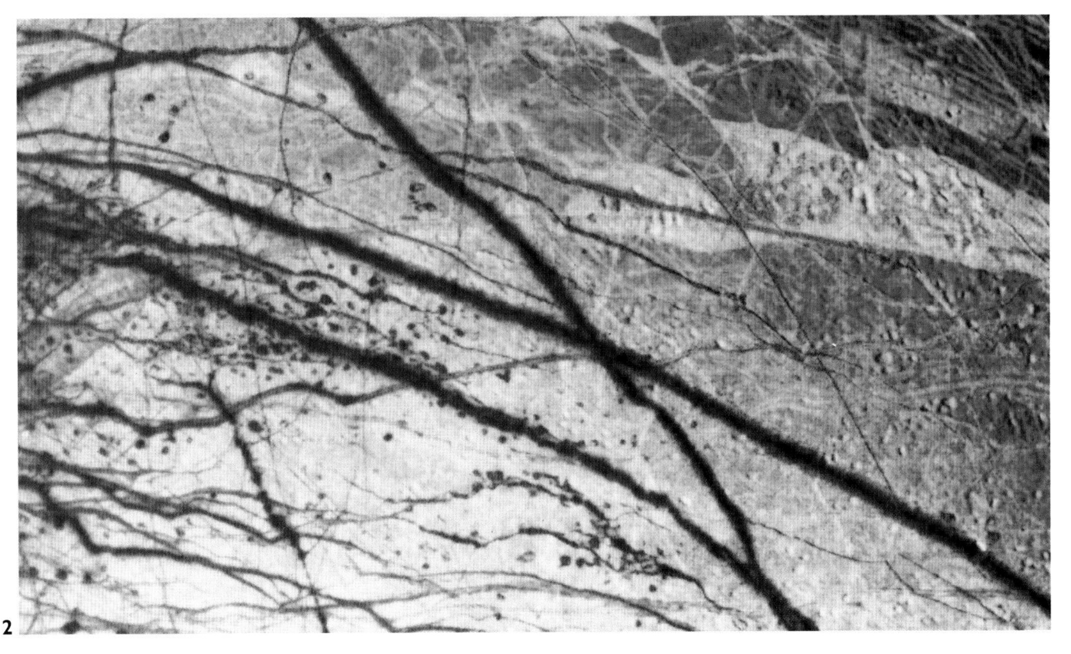

When pulsars were first discovered, it was thought that their regularly repeating 'signals' might be messages from little green men!

1. Family portrait of Jupiter and its four largest moons.

2. Close-up view of the icy surface of Europa, one of Jupiter's moons.

3. The life zone around the Sun.

of Europa. The result might not be comfortable for human beings, but if life can exist in the chilly waters of Antarctica, then in principle it can exist on Europa.

At once, the known comfort zone for life around our Sun doubled in size.

Heat from Inside

There are other ways to generate heat. The Earth itself is hot inside because of radioactivity, which keeps the core molten. If a planet like the Earth formed in the equivalent of an orbit between those of Jupiter and Mars, it might be too cold on its surface for liquid water to flow, but it could have very deep, frozen oceans, because very little of its primordial water would have evaporated. Any ice deeper than 14 kilometres below the surface would be melted by the internal heat of the planet.

At the end of the twentieth century, astronomers realized that they had been far too parochial in their estimates of the chance of finding liquid water elsewhere in the Universe and, therefore, in their estimates of the chance of finding life. But the Holy Grail still remains: what are the chances of finding other Earths?

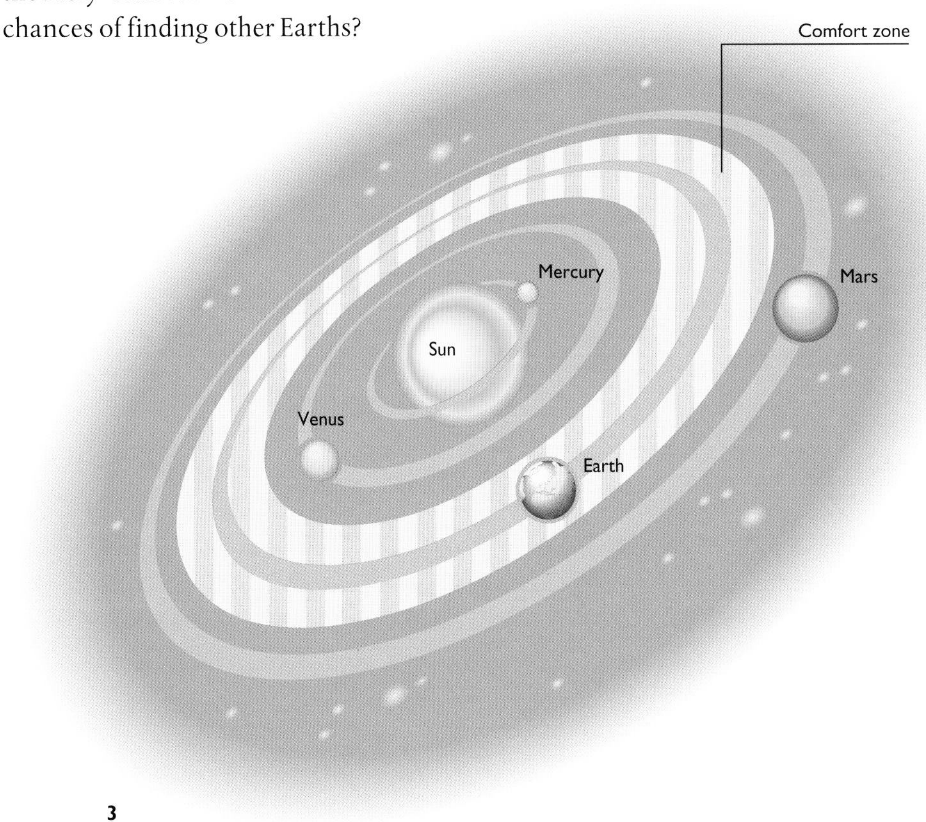

LIFE ON EARTH

One of the biggest mysteries in biology is how life got a grip on the Earth so soon after our planet formed. The Earth formed about 4.5 billion years ago, but it was bombarded with debris from space early in its life, and took about half a billion years to cool to the point where liquid water could flow. Yet there is fossil evidence of bacterial life in rocks at least 3.9 billion years old. Could life really have got started from scratch in just a hundred million years or so?

Cosmic Rain

Perhaps life didn't have to start from scratch on Earth. Although the bombardment of the Earth eased after about half a billion years, it didn't stop. Even today, the Earth is still occasionally struck by a large object (as the dinosaurs found to their cost 65 million years ago), and more gentle impacts with comets and smaller bits of cosmic debris happen quite often – on a geological timescale, that is. Since the 1960s, there has been a growing awareness that clouds of material in interstellar space contain a variety of complex carbon compounds, the raw materials of life. It now seems likely that cometary impacts when the Earth was young brought some of those raw materials down to the surface of our planet, giving life a kick start.

In 1986, cameras onboard the space probes Giotto and Vega sent back images of dark material coating the surface of the icy core of Halley's comet, and spectroscopy revealed that this is made of a variety of carbon-rich molecules. Ground-based telescopes have shown that the gases which make comets so spectacularly visible are also rich in carbon compounds, including methane and ethane. All this is important, because carbon is the key component of life (so much so that complex carbon chemistry is also called 'organic chemistry'). Microscopic particles of dust from space, much of it comet debris, are constantly falling onto the Earth, and analysis of samples collected at high altitudes tells us that about 30 tonnes of organic carbon compounds reach the surface of the Earth this way each day.

1

Seeds from the Clouds

The ultimate origin of all this stuff is the interstellar molecular cloud in which our Solar System was born. Spectroscopic studies have revealed the presence of many organic carbon compounds, such as formaldehyde, and ethanol (also known as vodka). But the most important discovery, made in 1994, was that of glycine, the first amino acid discovered in space. Out of more than a hundred molecules known to exist in space, this is the most important, because amino acids are the building blocks of proteins, and proteins are what your body is made of.

2

It is very hard to see how simple compounds like carbon dioxide and water could have developed into living bacteria in only a hundred million years. But if the young Earth were laced with complex organic molecules things like amino acids, then the whole process would have proceeded much faster.

The problem of the origin of complex organic molecules, the precursors of life, is moved from the Earth and out into space. There, although cold clouds of gas and dust may not seem the ideal places for chemical processes to take place, there was time enough – billions of years – for these first steps to be taken and for the seeds of life to form.

This would have happened before the stars and planets formed from the collapse of those clouds. The implication is that the 'seeds' of life fall upon every new planet, even if they do not always 'germinate' there.

3

1. Fossils reveal the history of life on Earth. The oldest fossils show that life on Earth existed almost as soon as the planet had cooled.

2. The horse-head nebula is a cloud of dust in space typical of the material from which planetary systems form.

3. The central core of Halley's comet, photographed by the space probe Giotto, is primordial material left over from the birth of the Solar System.

TOPIC LINKS

1.1 Stepping Stones to the Universe
p. 17 The Stars are Suns

1.2 Star Birth
p. 31 The Orion Nursery
p. 39 Dusty Discs

1.3 Stellar Evolution
p. 53 Cooking the Elements

3.2 Talking to the Stars
pp. 164–5 Panspermia

OTHER EARTHS?

The technique of detecting planets by the 'Doppler wobble' they produce in the light from their parent stars cannot yet detect the presence of Earth-like planets in Earth-like orbits around other stars. But there are two ways in which astronomers believe they may be able to detect such planets using ground-based telescopes – and other possibilities as space technology improves.

The first technique may already have worked. It depends on the way that light is bent when it passes near a massive object. This phenomenon is called 'gravitational lensing', and it was the technique used by astronomers in 1919 to test Albert Einstein's general theory of relativity. By studying the light from distant stars as it passed close by the Sun during a total solar eclipse, they found that it was bent by the Sun to exactly the degree predicted by Einstein's theory. The effect is bigger for a bending (or 'lensing') object of bigger mass, so it is much smaller for an Earth-sized planet than it is for the Sun. Nevertheless, it should be measurable even for planets orbiting other stars, if conditions are just right.

There are two ways in which conditions may be just right. First, when one star passes exactly in front of another star, as viewed from Earth, the lensing effect of the nearer star makes the more distant star seem brighter for several weeks. If the nearer star has an Earth-sized planet orbiting it, the planet may also pass in front of the more distant star, producing a smaller blip of extra brightness as it passes across our line of sight.

At the end of the 1990s, astronomers working at the University of Notre Dame, in Indiana, observed just such a double blip as one star passed in front of another. From the size of the smaller blip and its duration – just two and a half hours – they calculated that it could have been caused by a planet a few times more massive than the Earth, orbiting its star in the equivalent of the zone in the Solar System where Venus, Earth and Mars orbit our Sun. This single observation is not proof that other Earths exist, and there is no way to take a second look at this possible planet – it was a one-off event. But at least it shows that the technology is good enough for the job.

The Dragon Planet

Other astronomers have looked for the tiny increase in brightness that might be seen in a distant star when a planet in orbit around the star crosses in front of it. This would be a rare event, because there is no reason why the orbit of such a planet should be lined up with the Earth. The effect is also very small.But we would see a repeating pattern of tiny increases and decreases in brightness as the planet orbits round its parent star. Late in 1999, an international team of astronomers reported that they had found just such a pattern in the light from a system known as CM Draconis (which lies in the constellation of the Dragon). The pattern suggests that there is a planet with two and a half times the mass of the Earth orbiting in this system right in the comfort zone, where it receives roughly the same amount of energy that the Earth receives from the Sun. The discovery has yet to be confirmed, but it shows that the technology works, and we can expect more observations like these in the months and years ahead.

As the examples of Venus and Mars show, however, the presence of a rocky planet in (or near) the comfort zone doesn't prove that life exists on that planet. Are there any ways in which astronomers, exploring the Universe with their telescopes without ever leaving the Solar System, might be able to tell whether an object like the Dragon planet harboured life? Astonishingly, the answer is 'yes', and the technology to do the job might exist within 30 years.

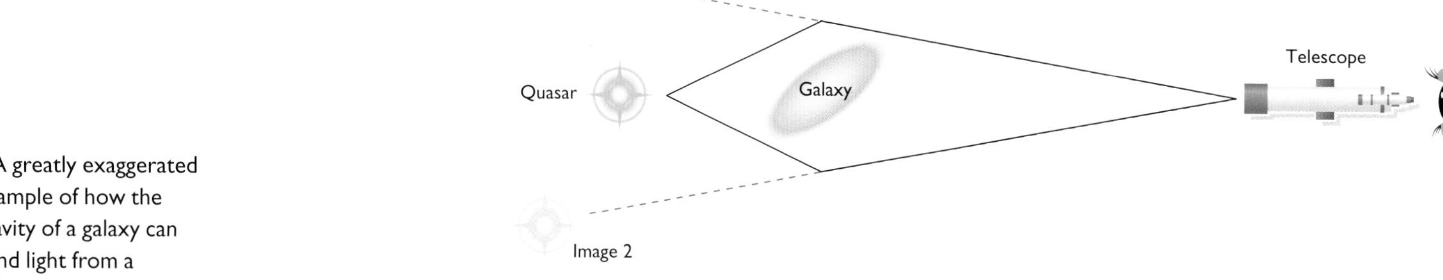

1. A greatly exaggerated example of how the gravity of a galaxy can bend light from a distant quasar.

2.The star 51 Pegasi, which is now known to have at least one planet in orbit around it.

The insight on which this possibility is based goes back to the 1960s, when the British scientist Jim Lovelock was designing instruments for NASA to use on space probes that would search for signs of life on Mars. Lovelock realized that this was a waste of time, because spectroscopy had already revealed that the atmosphere of Mars was made up of inert carbon dioxide, which meant that it was a dead planet. On Mars, the chemicals in the air are bound up in low energy states, and nothing interesting can happen to them (the same is true of Venus). On Earth, by contrast, there is a lot of highly reactive oxygen in the atmosphere, which is a mixture of gases in a high energy state, able to react with one another. If chemistry alone were at work, everything would react, producing an inert atmosphere. It is the presence of life, which uses sunlight to break up inert chemicals and release active chemicals to the air, which makes the Earth distinctive.

This insight led Lovelock to develop his idea of Gaia – the Earth as a single living system, with the physical environment and

A PULSAR WITH PLANETS

To the intense irritation of radio astronomers, what they regard as not only the first extra-solar planet to be discovered, but the first planetary system found outside the Solar System, is almost entirely disregarded by the astronomers who search for planets using optical telescopes. The discovery was made in 1992, from analysis of the radio pulses from a pulsar prosaically named PSR B1257+12. Using the radio equivalent of the Doppler wobble technique, the radio astronomers found the spectroscopic trace of not one but three planets in orbit around the pulsar. Strangely, as far as the orbits and masses of the planets are concerned, this is the most similar system yet found to our own Solar System.The main difference is that the planets orbit their star more closely than the three inner planets of our Solar System – but if the radii of their orbits were doubled they would be close to the orbits of Mercury, Venus and Earth.

The reason why most planet hunters ignore this discovery and regard the planet orbiting 51 Pegasi as the first 'real' discovery is that the planets orbiting PSR 1257+12 must have formed after the neutron star collapsed. Any original planets in the system would have been destroyed in the supernova in which the neutron star was born. So they must be very different objects from the planets in our Solar System, just as the pulsar is a star quite unlike our Sun. Nevertheless, they do orbit a star, and they are planets.

the biological environment interacting to maintain an essentially unstable balance. Forty years on, it has also led to the most promising way to search for life in space.

Seeking Signs of Life

The best way to detect Earth-sized planets will be to put large telescope arrays up into space. These will go far beyond the famous Hubble Space Telescope, both in terms of size and location. The idea is to put systems made up of several large dishes, linked together to act like one huge telescope, far away from the Earth where there will be no disturbances to disrupt their observations. Because planets are much cooler than stars, they radiate energy in the infrared part of the spectrum, with frequencies even lower (and therefore with less energy) than those of red light (▷ p. 18). These space telescopes are being designed to look at that part of the spectrum. If they find an infrared source alongside a star like the Sun, the chances are it will be a planet. But both the European Space Agency (ESA) and NASA hope to do better than that.

ESA already has such a system, called Project Darwin, on the drawing board. NASA's counterpart is the Terrestrial Planet Finder (TPF). It is likely that the two projects will be merged into one, to avoid costly duplication of effort. What makes them so attractive, in spite of the cost, is that the infrared part of the spectrum is just the place where characteristic signatures of chemical compounds associated with life on Earth occur. Oxygen itself, plus the tri-atomic form of oxygen known as ozone, and water vapour all leave their fingerprints in the infrared part of the spectrum. Telescopes like Darwin/TPF could be flown in space as early as the 2030s, and would immediately be able to detect planets in orbit around stars 3 to 5 parsecs away. Obtaining spectra may take a little longer, but if these projects go ahead we should know for sure, within 50 years, whether these planets are dead, like Venus or Mars, or whether they really are other Earths.

1

ANOTHER 'SOLAR SYSTEM'

In 1997, astronomers discovered a planet with roughly the same mass as Jupiter in orbit around a star in the constellation Andromeda, using the Doppler wobble technique. Two years later, they discovered that there are two more giant planets orbiting the same star. It took so long to work out that there were actually three planets in this system, and to determine their masses and some details of the kind of orbits they are in, because the spectroscopic signatures of the planets in the light from the star interfere with each other. But by the spring of 1999 it was clear that the complicated pattern of Doppler shifts being seen in the light from this star could most simply be explained as the presence of three planets, not one.

The mass of the star is about 30 per cent bigger than that of the Sun, and it is about halfway through its lifetime on the main sequence (▷ p. 51), which will be about six billion years. The system is 13 parsecs away from us. The innermost of the three known planets zips round the star once every 4.6 days. It has about 70 per cent of the mass of Jupiter. The next planet out from the star has a year 242 Earth days long and is twice the mass of Jupiter, while the third takes 1270 Earth days to go round its star once and has four times the mass of Jupiter. There is nothing unusual about either the star or its planets, and this is the first proof that systems of planets, instead of just individual planets, exist outside the Solar System.

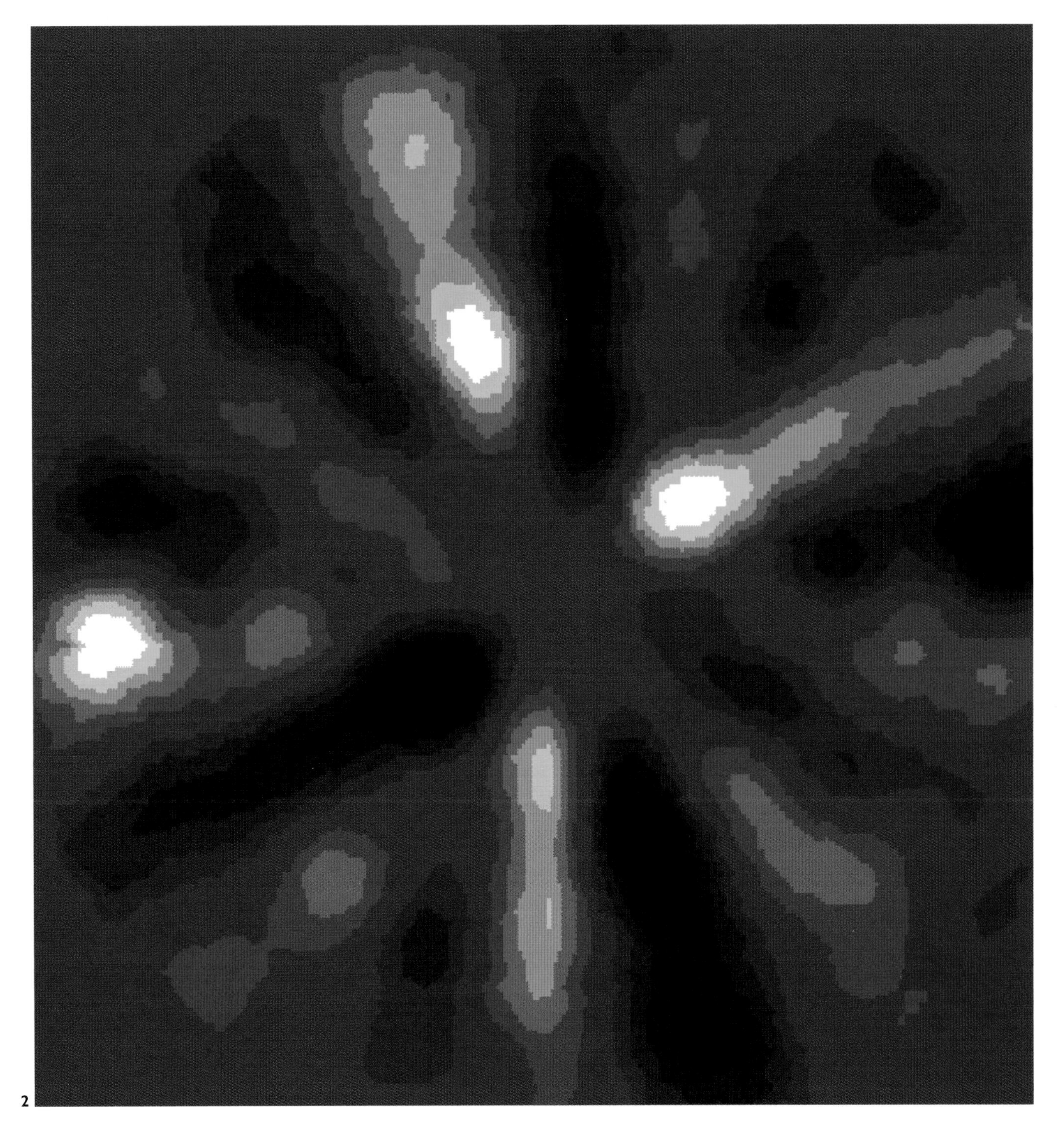
2

1. Artist's impression of the kind of space telescope needed to obtain images of planets orbiting other stars.

2. A simulation of how the inner part of our Solar System would look to a telescope like the Darwin system.

3.2 TALKING TO THE STARS

People interested in making contact with other civilizations beyond our Solar System usually talk of the 'Search for Extra Terrestrial Intelligence', or SETI. They assume that we will be able to make contact with such intelligences, if they exist, either by radio communication or by sending space probes to the stars. But what those astronomers are really looking for is evidence of extraterrestrial technologies. Of course, we would have no hope (yet) of detecting a Roman-level civilization on an Earth-like planet orbiting another star, even though the Romans of the first century AD were both intelligent and civilized (in their way). And there may be civilizations that are in some sense more advanced than ours – older and more peaceful, perhaps – that never developed machines and radio communication. But for now, such possibilities remain in the realms of science fiction; the only real prospect of finding extraterrestrial intelligence is by using technology – in particular, one specific piece of technology, radio.

FIRST STEPS

Politicians (and other people) sometimes express concern about the possibility of making two-way contact with alien intelligences. They fear that the experience of encountering a superior civilization might destroy the civilization we already have here on Earth, in much the same way contact with Europeans destroyed the culture of the native Americans. It was for this reason that astronomers hoping to make contact (even if only one-way contact) changed the name of their programme from its original CETI (Communicating with ExtraTerrestrial Intelligence) to SETI, implying that we could listen out for other civilizations without necessarily revealing our presence.

Making Contact by Mistake

But it is already too late – we have been broadcasting to the Universe, more or less indiscriminately, for more than half a century. The first radio broadcasts could not escape into space, because they were reflected back from the ionosphere, a layer of charged particles high in the Earth's atmosphere (they were also very feeble, and would be difficult to detect anyway at stellar distances). The first signals that had both the right wavelength to penetrate the ionosphere and the power to be detected at reasonable distances were the broadcasts made by the BBC from the 1936 Olympics, in Berlin. These were quickly followed into space by early television programmes, radar pulses, and more recently by the powerful radio signals beamed to interplanetary space probes to control their activity.

1. TV camera photographed in 1936 in front of the transmitting mast of the Alexandra Palace.

1. Star cluster in a large magellanic cloud, taken by the Hubble Space telescope.

2. We've already begun broadcasting to the Universe. Given sufficiently advanced technology, aliens in distant star clusters will one day be able to watch *Sergeant Bilko*.

1

The bowl of the Arecibo radio telescope is big enough to hold four billion bottles of beer.

All of these signals travel at the speed of light, in different directions out through space. So the Solar System sits in the middle of an expanding bubble of radio noise (for these purposes, short wavelength electromagnetic radiation such as TV and radar signals are classed as radio waves), extending in all directions out to a distance of about 20 parsecs, and getting bigger by 0.5 parsecs every year. The signals at the outer edge of the bubble are pretty feeble; but the ones following the radio waves from the 1936 Olympics get stronger and stronger as the technology of broadcasting has improved.

This bubble has already spread beyond

scores of stars – there are a couple of dozen within 3 parsecs of the Sun, which have already had the opportunity to experience the Berlin Olympics, *I Love Lucy* (in its original broadcasts), *Monty Python's Flying Circus*, and CNN's coverage of the Gulf War. Because of the time taken for the signals to travel such huge distances, these stars still have the global broadcasts celebrating the start of the third millennium to look forward to.

It is quite possible that our first contact with an alien civilization will come in the form of a communication beamed directly at our Solar System in response to the radio noise we are already making – perhaps even a polite request to turn the noise down a bit.

Shouting at the Stars

There have also been deliberate attempts to attract attention to ourselves – the radio equivalent of shouting, as loud as possible, 'Here we are'. The best way to do this is to use the largest radio telescope in the world, which has been built in a natural hollow in the ground at Arecibo, in Puerto Rico. Because it is built in a bowl in the ground, the Arecibo Radio Telescope can only cover a strip of the sky as the Earth rotates; it cannot be steered to point in other directions. However, it has a diameter of 305 metres, giving it a collecting area bigger than every optical telescope built in the twentieth century put together. This helps it to detect very faint radio noise from space, provided it comes from the right part of the sky. It also makes it ideal (equipped with a suitable broadcast system) for sending a powerful beam of radio waves out into space. In this role, it has been used to bounce radar pulses off Venus and Mars and detect the

2

ZAPPING THE STARS

Laser beams, which use light instead of radio waves, unlike the Arecibo Space Telescope (right), have much shorter wavelengths and only spread out a little (only enough to cover a whole planetary system) even on a journey tens of parsecs long. The snag is that you can only aim at one star at a time, so you have to know there are planets there before you send out your beam.

At a distance of parsecs, or even tens of parsecs, the light from a powerful laser (broadcast from a space station orbiting the parent star) would merge with the starlight, and would not be detectable as a separate spot of light. But because a laser beam is essentially an intense beam of light all with the same wavelength, any observers with technology as good as our own would see it as a very peculiar feature in the spectrum of the star, shifting to and fro with a distinctive Doppler rhythm. By turning the laser on and off, we could even send messages in binary code. Such a system could be detected by us from a distance of 3 parsecs or more.

The laser required for such a project need not even be very powerful. A few tens of kilowatts would do, provided it is mounted in space, above the obscuring layer of the Earth's atmosphere.

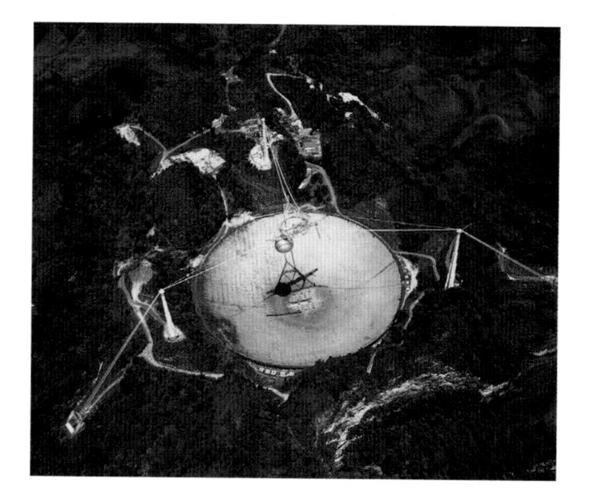

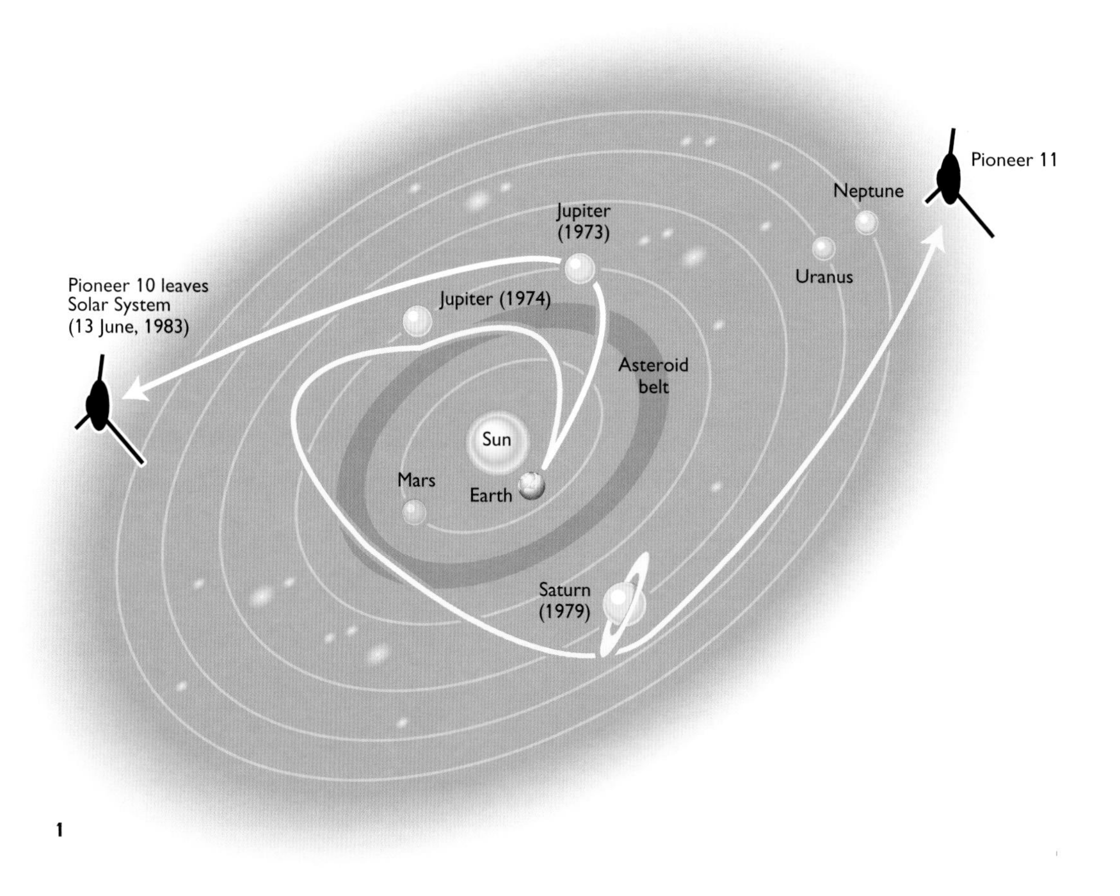

1 and 2. The identical space probes Pioneer 10 and Pioneer 11 were the first artificial objects to leave the Solar System.

reflections, measuring distances across the Solar System with unprecedented accuracy.

In 1974, astronomers from Cornell University used the Arecibo Radio Telescope to broadcast the first deliberate message sent by human beings to other stars. This blast of radio waves was beamed in one direction, and will spread out slightly on its way through space. It will not have been detectable from planets orbiting most of the stars in the local volume of space already polluted by the Berlin Olympics, *I Love Lucy* and the rest. In order to give this relatively narrow beam the best possible chance of being picked up by another technological civilization, the astronomers directed it towards a globular cluster of stars, containing 300,000 individual stars, in the direction of (but far beyond) the constellation Hercules. If there are civilizations capable of detecting radio waves on any planets orbiting any of those 300,000 stars, they could pick up this message, and this hugely increases the chance of a response.

But there's a snag: the cluster actually lies about 740 parsecs away and radio signals travel at the speed of light (there and back). If a civilization in the cluster does pick up the signal from 1974, and responds immediately, we can expect a reply in AD 50,000.

SIGNALLING TO THE UNIVERSE

But how can we hope to communicate meaningfully with alien civilizations, rather than just making a noise? The scientists involved in sending the 1974 signal from Arecibo wanted to send a message that contained information that an alien might be able to decipher, so they used what is considered the universal language – mathematics. One of the scientists involved, Carl Sagan, described the message in words at the time:

'What it said fundamentally was: "Here's the Sun. The Sun has planets. This is the third planet. We come from the third planet. Who are we? Here is a stick diagram of what we look like, how tall we are, and something about what we're made of. There's four point something billion of us, and this message is sent to you courtesy of the Arecibo telescope, 305 metres in diameter."'

It turns out to be surprisingly easy to convey that kind of information in the simplest mathematical language of all – the binary 'on-off' code familiar from computers. Frank Drake, a keen enthusiast for SETI (see later), tested this in the 1960s by devising just such a message (later used as the basis for the Arecibo message) and sending it out to colleagues to see if they could crack the code.

Drake's Message

Drake's message consisted simply of a string of 0s and 1s – binary code – 551 characters (or 551 'bits' in computer language) long. Any mathematician would quickly realize that

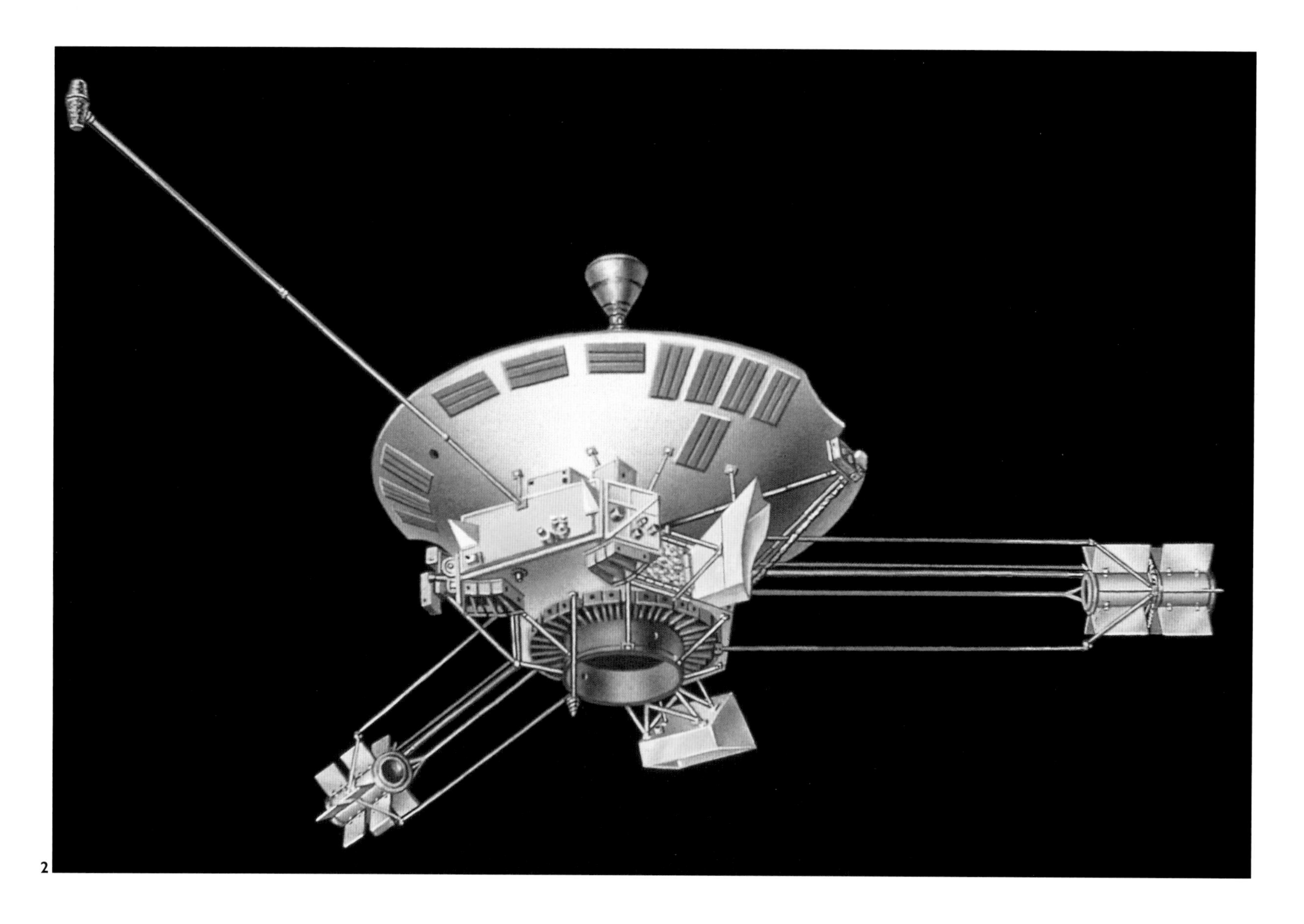

2

The name 'nano', as in 'nanobacteria' and 'nanometre', comes from the Greek word *nannos*, meaning a mischievous dwarf.

551 is a product of two prime numbers, 19 and 29. The only two numbers that multiply to give 551 are 19 and 29. But 19 and 29 do not divide by any whole number. This suggests (to the mathematically inclined) that the string of 0s and 1s can be turned into a rectangular 'picture' in either of two ways – a grid with 19 rows of 29 characters, or a grid with 29 rows of 19 characters. Trial and error would quickly show that the first grid is gibberish, but the second produces a distinct pattern if you put a black square everywhere there is a 1 and a white square where there is a 0 (or vice versa).

Drake made up his message as if it came from an alien civilization, so the pattern he made represents the imaginary civilization. The little picture of the alien itself is fairly obvious. Down the left-hand side of the picture there is a representation of a star and its nine planets, with a rough indication of their sizes. The other bits of code are mostly numbers. The numbers 1 to 5, written in binary, alongside the first five planets, then 11 alongside planet two, 3000 alongside planet three, and 7 billion, alongside planet four. Drake intended this to mean that 7 billion aliens lived on planet four, they had a colony on planet three, and that at the time they sent the message there was a small expedition on planet two. The diagrams at the top right of the picture represent the atoms carbon and oxygen, telling us a little about the aliens' chemistry.

No individual scientist managed to decipher all of Drake's 'alien message' (by the way, he didn't pretend it really did come from aliens; they knew he had made it up as a test). But any real message from the stars would be investigated intensively by high-powered teams of scientists from all disciplines and they would be able to crack any similar code. This shows what an impressive amount of information can be packed into just 551 bits of computer code. With 8 bits making a byte, that is less than 70 bytes of information, and computer memory these days is measured in megabytes and gigabytes.

1

Messages in Cosmic Bottles

You may think that radio waves take a long time to get to the stars, but space probes, crawling along at a tiny fraction of the speed of light, take vastly longer. Even so, space scientists have already sent messages beyond the Solar System attached to space probes. These were not probes designed for the purpose, but ones that happened to be heading out of the Solar System anyway, after completing their missions of exploration of the outer part of the Solar System. This is the equivalent of throwing a message in a bottle into the ocean, and hoping that it might be picked up by a passing ship; a long shot, but the

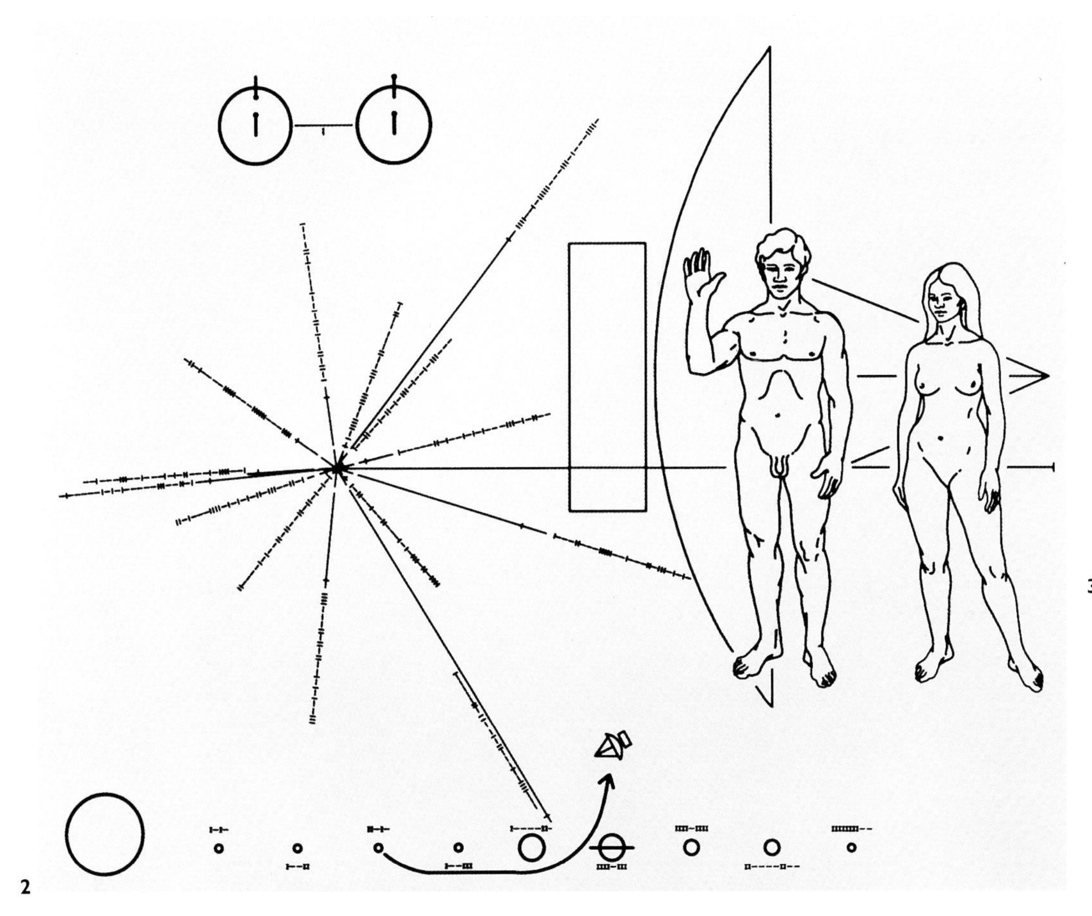

1. Frank Drake, with his eponymous equation.

2 and 3. The two Pioneer space probes which flew past Jupiter on their way out of the Solar System, each carried a plaque with scientific information and representations of a man and a woman.

opportunity provided seemed too good to miss.

The spacecraft Pioneer 10 was the first human artefact to leave the Solar System. Launched in 1972, Pioneer 10 and its twin probe Pioneer 11 each flew past Jupiter and then followed slightly different trajectories to provide, between them, our first close-up views of all the gas giant planets. The landmark date was 13 June 1983, when Pioneer 10 crossed the orbit of Neptune and officially left the Solar System (Pluto, which is usually the furthest planet from the Sun, occasionally dives just inside Neptune's orbit, and did so between 1979 and 1999). It became the first message in a bottle from the human race to be launched into the ocean of interstellar space.

Each of the Pioneer 10 and 11 spacecraft carried an identical plaque, designed by Carl Sagan and Frank Drake, as a 'hello' to any intelligent beings that might find it. The plaque includes a representation of the spaceprobe itself, the Sun and Solar System, and a 'map' indicating the location of the Sun relative to some of the pulsars. It also includes a rather stylized drawing of a naked man and woman, which drew an angry response from some American citizens concerned that NASA was polluting the Universe with pornography.

When the next missions to Jupiter and beyond were launched in 1977, they carried a more sophisticated message, in the form of a video disc. Each of the two Voyager craft carried a copy of the disc, sealed in a container engraved with scientific information including instructions on how to play the disc. The disc itself contained images of the Earth, sounds ranging from whale song to Chuck Berry, scientific information, and messages from UN Secretary General Kurt Waldheim and US President Jimmy Carter. In truth, any alien who finds the disc is unlikely to learn any more from its contents than the fact that we exist. All subsequent missions to the outer planets have gone into orbit around them, and only Pioneers 10 and 11 and Voyagers 1 and 2 have left the Solar System (▷p. 156).

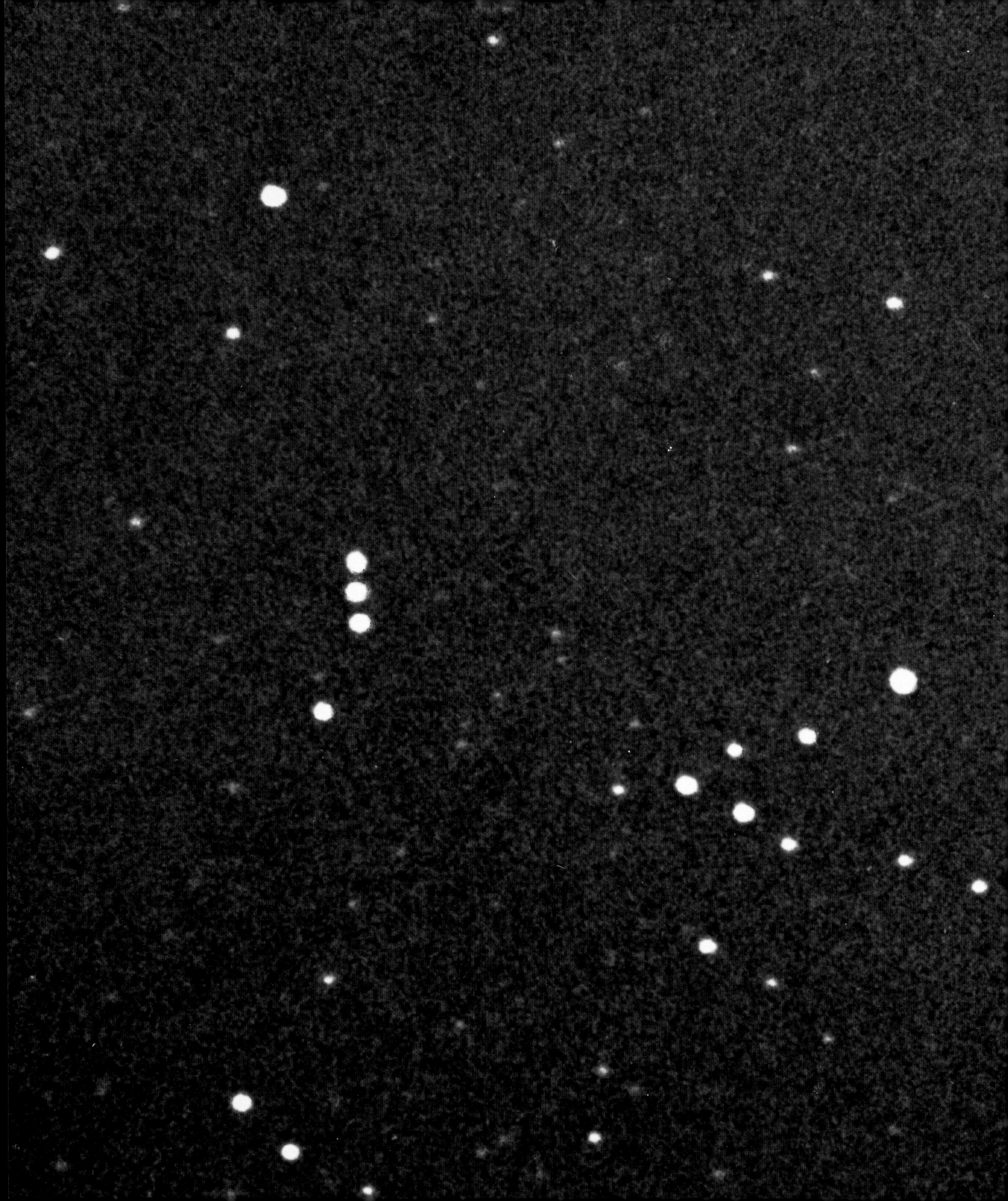

TO BOLDLY GO

The Pioneer and Voyager probes are not heading towards any nearby star. Their trajectories out of the Solar System are a result of their encounters with the giant planets, and were not chosen with interstellar travel in mind. But if a probe were to be sent on a voyage of exploration to another star, the obvious place to go would be to the nearest star, Proxima Centauri. Even Proxima, though, is more than a parsec from our Solar System, and some ingenuity (plus considerable financial input) will be needed to send a probe there and get back data. There are two possible approaches that could be put into practice in the twenty-first century if anyone has the inclination and the finance to do it.

Brute Force

The brute force approach is typified by plans drawn up by the British Interplanetary Society (BIS). The BIS carried out a five-year feasibility study of an interstellar probe powered by nuclear fusion. The design calls for the kind of spaceship familiar from fiction, 200 metres long and with a mass of 54,000 tonnes, assembled in orbit around one of the moons of Jupiter. The fuel for the probe is envisaged as compressed pellets containing the heavy form of hydrogen (deuterium) and the light form of helium (helium-3). The design calls for pellets to be injected into a fusion chamber at a rate of 250 per second, where they would be zapped by beams of electrons, causing them to fuse and release energy in much the same way that nuclear fusion releases energy at the heart of a main sequence star. The rate of the explosions, 250 per second, is roughly the same as the rate at which individual petrol vapour explosions occur in the cylinders of a car while it is quietly ticking over.

Like the big rockets that launch payloads into space from Earth, the BIS design, dubbed *Daedalus*, is in separate stages, so that one can be discarded when its fuel is exhausted. The first stage of *Daedalus* is designed to run for two years before being dropped behind, with a second stage

1. (opposite) Three superimposed photos taken over a period of four years (the three images forming a dotted line) show Barnard's star moving relative to more distant stars.

2. A representation of the starship *Daedalus*, designed by the British Interplanetary Society.

1. Hypothetical depiction of a planet (foreground) orbiting the star Proxima Centauri.

operating for 22 months before the fuel is exhausted. All that effort will have pushed the speed of the 400 tonne payload to about 13 per cent of the speed of light. The target suggested by the BIS for the hypothetical *Daedalus* voyage (because it looks more interesting than Proxima) is Barnard's star, about2 parsecs away. The journey would take about 50 years, Earth time, and then *Daedalus*, with no means of slowing down, would hurtle through the Barnard's star system in about 20 hours, observing and recording data to send back at leisure as it continued on its way into the Universe.

Nobody expects that the first spaceprobe will be exactly like *Daedalus*. But the existence of this design study shows what could be achieved using what you might call the conventional approach.

Sailing to the Stars

There is an alternative. American space scientist Robert Forward is the leading proponent of a more subtle approach to exploring the cosmos – 'starsailing'. The problem with rockets, as Daedelus shows, is the enormous weight of fuel required – 54,000 tonnes of rocket to deliver a 400 tonne payload (less than 1 per cent of the original mass of the spaceship). Forward's solution is to leave the power system behind and only send the payload to the stars. He has designed a system that could get a 1-tonne payload to Proxima Centauri or its near neighbour Alpha Centauri in less than 40 years. It does so using a thin 'sail' 3.6 km (2 miles) in diameter, but only 16 nanometers (16 billionths of a metre) thick, made of aluminium. Unfurled in space and attached to the probe at its centre, the sail would be sent on its way out of the Solar System by the beam from a huge laser, also built in space. The laser would give the sail an acceleration of 3.6 per cent of the gravitational force we feel at the surface of the Earth, sending it up to a speed of about one-tenth of the speed of light (at this speed, it would take two days to travel from Pluto to the Sun).

As with *Daedalus*, Forward's probe, which he calls *Dragonfly*, could be left to hurtle through the target system, sending back data. Supporters of this approach to interstellar travel point out that if there are any beings worth talking to in the target systems, they will certainly notice it coming and they may well be able to stop it for themselves.

Starsailing is currently the hot topic in interstellar probe design, with many variations on the theme being developed by Forward and his colleagues. It offers a realistic prospect that a probe could be on its way to the nearest star before the middle of the present century and could be sending back data from its target star before the end of the century. Very possibly, babies already born will still be alive when news from Alpha or Proxima Centauri gets back to Earth.

TIME DILATION

If we ever do send space probes to the stars, we will see an effect predicted by the special theory of relativity at work. The effect is called 'time dilation', because the clocks on a spaceship, moving at a sizeable fraction of the speed of light, will run slow compared with clocks at home. Time literally passes more slowly on the moving spaceship than it does on Earth. The effect only really shows up at very high speeds, but it would already be detectable for a probe like *Daedalus* or *Dragonfly* (right), travelling at about 10 per cent of the speed of light.

The correction factor required is called the Lorentz factor after the Dutch physicist Hendrik Lorentz, who worked on the theory. At 10 per cent of the speed of light, the Lorentz factor is 1.02, which means that for every hour that passes on the spaceship, 1.02 hours (61.2 minutes) pass on Earth. If the journey seems to take 40 years to us, to the spaceship it takes only 39.2 years.

The Lorentz factor gets bigger the faster you go. At half the speed of light, it is 1.15, so that 15 per cent less time elapses for the voyagers than for stay-at-homes. If you could travel fast enough, this would help to make such long voyages more practicable, but the energy required to make the ship go fast enough to make this worthwhile is prohibitive.

If you could travel at half the speed of light, however, and went on a round trip voyage (there and back) lasting 50 years by your spaceship clocks, you would be 50 years older but you would find everything on Earth 57.5 years older than when you left. The effect is real and provides a kind of one-way time travel, into the future.

PANSPERMIA

Human civilization stands on the brink of being able to 'seed' the Universe deliberately. We have already sent four spaceprobes beyond the confines of the Solar System. But, as is often the case, anything people can do Nature can do better. It may be that we owe our own existence to interstellar travellers – not little green men, but bacteria sealed inside grains of cosmic dust.

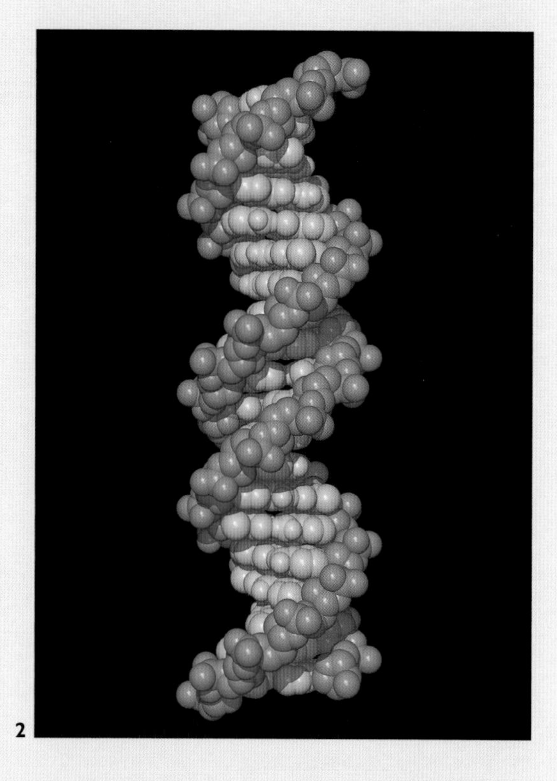

2

Life is ubiquitous on Earth, and grains of material containing bacteria even float high in the atmosphere of our planet. It is conceivable that such grains could escape into space today, and even be blown right out of the Solar System by the 'wind' of particles and radiation from the surface of the Sun. But, if they did so the DNA of any living things they contained would be destroyed by solar radiation and cosmic rays. Heavier grains of material could shield living microorganisms from the radiation, but such relatively heavy grains could not travel far from Earth, at least, not today.

When the Sun reaches the next phase of its life and swells up to become a red giant, however the amount of energy being radiated from its surface will increase hugely. This will help to blow away a great deal of material from the atmosphere of the Sun into space. It will mix with the material of the interstellar medium to form new molecular clouds. At this stage of its life, the Sun would easily expel material from the inner part of the Solar System, including life-bearing grains of dust from the Earth.

1

When bacteria are exposed to an extreme environment (particularly one without any water) they are capable of going into a form of hibernation, preserving their core of DNA and other life molecules until warmth and water are available once again.

Everywhere that life exists on planets around Sun-like stars, living material, sheltered inside grains of solid material, will be expelled in this way as the parent star becomes a red giant. It will then lace the molecular clouds from which new planetary systems form. When new stars and planets do form from the collapse of the molecular cloud, the comets that become part of those systems will also contain fragments of living (or once living) material. As the new planets cool, comets and dust from the proto-planetary disc will carry this material, including DNA, and possibly living organisms, down to their surfaces.

This idea, known as panspermia, is the best explanation of how life got a grip on the surface of the Earth so soon after the planet formed. As the proponents of panspermia point out, it doesn't even matter if the living bacteria are killed by the harsh conditions of interstellar space. Even fragments of DNA arriving in the oceans of a young planet like the Earth will give life a head start. If they are correct, the implication is that all of life on Earth is descended from the DNA of cosmic bacteria. It would be quite easy to test the idea by sending a probe to a comet and bringing back samples of its material to investigate for traces of DNA. If the panspermia hypothesis is correct, there should be biological material very similar to that of life on Earth in these fragments. And that would also mean that life elsewhere in the Milky Way Galaxy must be based on the same kind of DNA that we are, although it would undoubtedly have evolved into other forms as superficially different from one another, and from us, as a peony is from a pony.

It would also mean that life is common in the Milky Way, and that the effort of trying to make contact is likely to prove worthwhile.

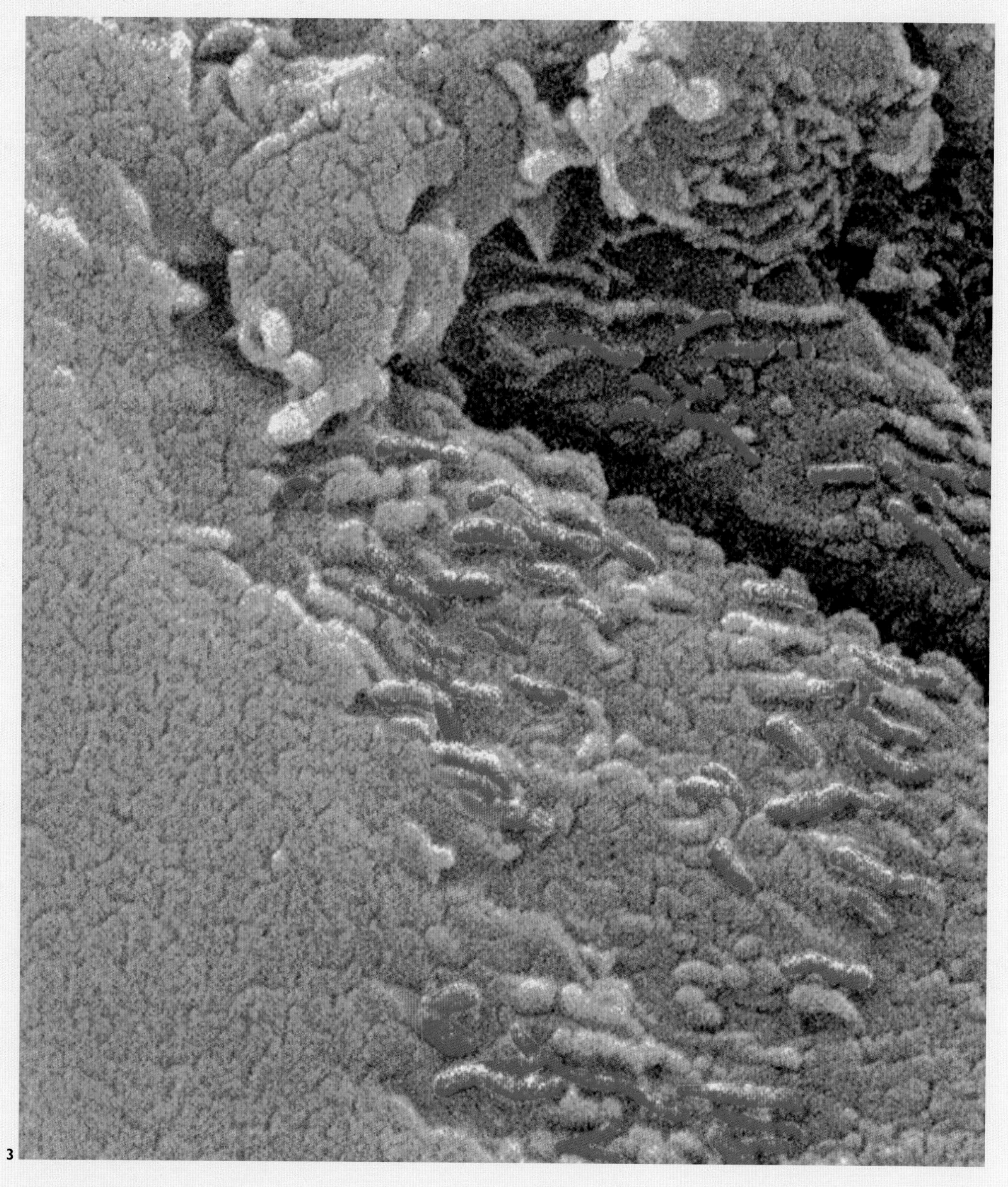

3

1. If the idea of panspermia is correct, comets like Halle-Bopp, pictured here in 1997, could have brought the seeds of life to Earth.

2. Model of DNA, the molecule of life.

3. Life on Mars? According to some researchers, the tube-like structures shown here, found in a meteorite which originated from Mars, are microfossils of bacteria-like organisms.

TOPIC LINKS

1.1 Stepping Stones to the Universe
p. 22 Stars in Spiral Patterns

1.2 Star Birth
p. 34 Dusty Beginnings

3.1 Life and the Universe
pp. 146–7 Life on Earth

3.3 LISTENING TO THE STARS

From the time when the technology to transmit and receive radio signals was developed (more or less at the end of the nineteenth century) to the time when live TV transmissions were made from the first astronauts to reach the Moon to anyone on Earth who had the right equipment to receive them was less than 70 years – less than a single human lifetime.

Once a technological civilization invents, or discovers, radio, it is likely to move on rapidly to cheap, powerful radio transmissions that are capable of being detected by comparable civilizations at distances of tens of parsecs. To many people this suggests that it makes more sense for us to concentrate on developing technology to listen for alien broadcasts rather than trying to signal to them ourselves. Such efforts to detect extraterrestrial intelligence are already underway.

Previous page. Optical image of comet Hale-Bopp showing both its gas and dust tails. The gas or 'ion' tail (blue) consists of gas blown away from the comet head by a solar wind.

THE INFINITE RADIO SET

Although radio is likely to be invented when a technological civilization is young, a more advanced civilization will still know about radio waves. They will be aware that this is a good way to communicate with people like us, even if they have found better ways to communicate among themselves. The best way to make contact with a newly emerging technological civilization (which, for all our achievements to date, is what we are) is by radio. This is the reason why the first generation of searches for extraterrestrial intelligence, roughly from the early 1960s to the end of the twentieth century, concentrated on the search for radio signals. But where do you search?

The first thing to decide when carrying out a search for extraterrestrial intelligence by listening out for their radio signals is which radio frequencies to listen at.

The important point about all SETI programmes carried out so far is that they are looking for deliberate signals – beacons of some kind designed to attract attention. Our

1. Radio astronomy observatories such as the Very Large Array link several radio telescopes together to mimic the power of a much larger instrument.

technology is not yet good enough to be able to eavesdrop on communications that are not intended to be picked up by people like us – the alien equivalent of *Monty Python*, perhaps. At first sight the search is a daunting task, because the aliens could be transmitting on any one of an infinite number of radio frequencies. It's no good if our radio receivers are tuned in to the cosmic equivalent of Radio 1 if the aliens are transmitting on Radio 4. We'd never hear them. But nature does impose some limits on the possibilities, and astronomers have also tried to get inside the minds of the aliens to work out logically the frequencies most likely they're broadcast on.

Limiting the Options

The first limitation is imposed by the atmosphere of the Earth. The atmosphere blocks out radio waves with frequencies outside the range from about 1000 megahertz (MHz) to 10,000 MHz. We physically cannot listen outside this range (except with instruments flown in space, above the obscuring layer of atmosphere, and no SETI project has yet gone that far).

The first and most important thing that astronomers learned about the Milky Way using radio telescopes is that it is full of hydrogen gas. This gas radiates radio noise at a frequency of 1420 MHz (equivalent to a wavelength of 21 cm). This enabled radio astronomers to map the spiral arms of our Galaxy by looking at where the hydrogen clouds emitting this radio noise were located.

1. The 43-diameter dish antenna of the Green Bank Observatories Radio Telescope.

2. A map of the Milky Way made by radio telescopes studying emission from hydrogen gas.

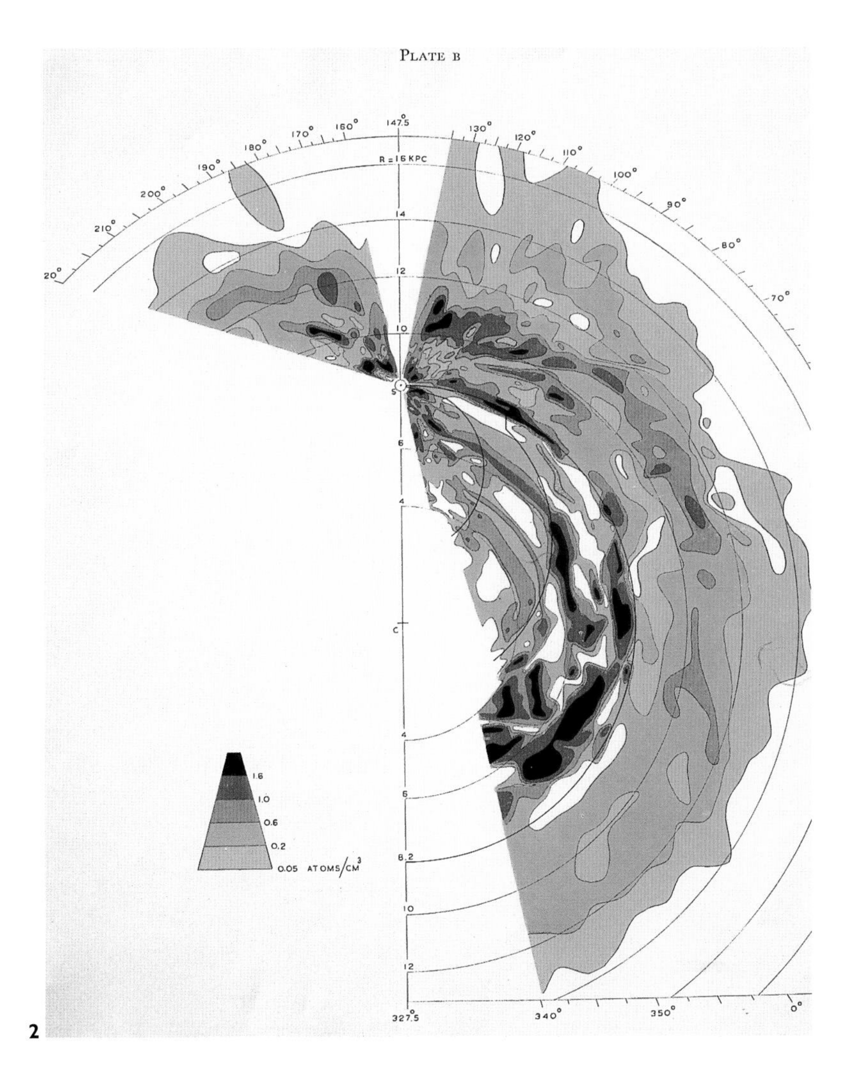

2

The cost of sending an interstellar message using existing radio telescopes is about $1 per word. The potential return on this investment is incalculable.

Any other radio astronomers in the Galaxy would also have found this frequency invaluable in their work, and would have developed sensitive receivers designed to detect the '21-cm' radiation. Anybody who investigates the nature of the Milky Way must have sensitive receivers capable of detecting the 21-cm radiation, and would realize that other radio astronomers would also have such receivers. So, the argument runs, alien scientists would guess that civilizations like us possessed equivalent detectors, and would therefore broadcast on a frequency close to the natural frequency of hydrogen. For this reason, the first SETI programmes were carried out using more or less standard radio telescopes, designed to pick up signals near 1420 MHz.

The pioneering example of this approach, headed by Frank Drake of Cornell University, was Project Ozma, named after the Queen in the fictional land of Oz. It looked for signals from several nearby stars in the early 1960s, but without success. Nevertheless, Project Ozma was the inspiration for many other searches that followed in the 1960s and 1970s. In one of these surveys, carried out in the early 1970s using a radio telescope (actually two different antennae) at Green Bank, West Virginia, ten stars were selected for investigation. One or the other of the antennae of the radio telescope was pointed towards one of the stars whenever this could be fitted into the regular research programme (which actually did involve mapping hydrogen clouds in the Galaxy). At the end of the survey, the team concluded that the study had not revealed the presence of any alien signals, and that substantial improvements in the number of stars studied, the area of sky searched, the sensitivity of the apparatus and the range of frequencies searched would be needed to make SETI worthwhile. Since then, progress has been made on all those fronts.

APPLYING NEW TECHNOLOGY

From the early 1980s onwards, the ability of radio astronomers to listen out at different frequencies was greatly increased by the huge advances in computer power and miniaturization. This made it possible to design compact, relatively simple systems that could automatically search a range of frequencies, looking for the kind of orderly signals that might be produced by intelligent beings. One reason this was an important step forward is that even if aliens were broadcasting at precisely 1420 MHz (21 cm), the signal we receive would be shifted slightly by the Doppler effect (▷ p. 82). In fact, it would be shifted by a combination of Doppler effects caused by the motion of the aliens' planet around their home star, the motion of that star and the Sun relative to one another through space, and the motion of the Earth around the Sun. But in the 1970s, astronomers had also come up with another bright idea to extend their search for extra-terrestrial life, and this too was just right for application of the new technology.

The Best Place to Listen

The region of the electromagnetic spectrum around the 21 cm wavelength is ideal for us to listen on, quite apart from the presence there of the important radio emission from hydrogen. At longer wavelengths (corresponding to lower frequencies), there is a great deal of radio noise, produced by free electrons roaming through interstellar space. At shorter wavelengths (corresponding to higher frequencies), it requires a huge amount of energy to transmit any intelligible information, so it is unlikely that any aliens will be broadcasting in that part of the spectrum. But at so-called 'centimetre wavelengths' (which includes the region from 20 to 30 cm), the Universe is both quiet and dark, so that even a relatively weak signal from another civilization could be heard above the background noise, the hiss of cosmic static that fills the radio band of the Universe.

This may seem surprising, because clouds of hydrogen gas 'broadcast' in exactly this part of the spectrum. But those clouds are

LIFE AS WE DON'T KNOW IT

One of the most important scientific discoveries of the 1990s was that the interior of the Earth – a dark place where there is no sunlight – is teeming with tiny life forms called nanobacteria (right). The total mass of all these living cells is thought to outweigh the mass of all the multi-celled life forms that we are familiar with. Some of these 'bugs' are living creatures only a few tens of nanometres in size (a nanometre is a billionth of a metre), but there are an awful lot of them.

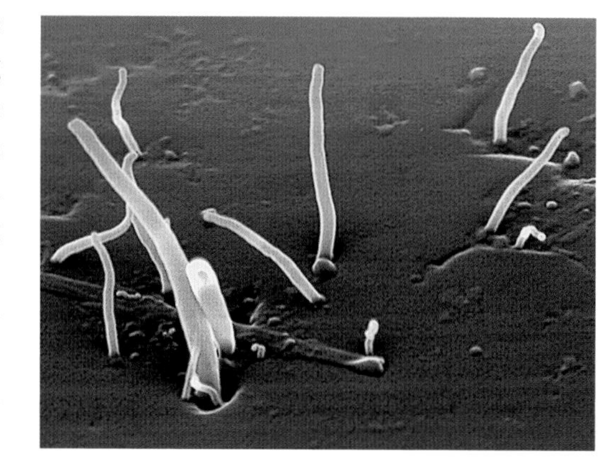

The discovery was doubly exciting because at about the same time other scientists identified fossil traces of what might be similar life forms in fragments from a meteorite which is thought to have travelled across space from Mars to the Earth. Perhaps there really is life on Mars (or in Mars!), although not 'life as we know it'. Indeed, some people argue that nanobacteria, safely tucked away inside the rocks of a planet, may be the most common form of life in the Universe.

Unfortunately, we cannot communicate with tiny bugs buried deep beneath the surface of a planet, and efforts to communicate with other intelligences in the Universe will have to focus, for the foreseeable future, on attempting to contact technological civilizations similar to our own. But it is worth remembering how restrictive this natural concentration on life as we know it may be.

1. Animals meeting at a waterhole. The cosmic equivalent may provide a meeting place for alien intelligences.

very cold and the radio noise they broadcast is quite feeble. It is only because this part of the spectrum is indeed so quiet that we are able to detect them at all, let alone use them to map the structure of the Galaxy. Any civilization just a tiny bit more advanced than our own would have no trouble 'shouting' loud enough to be heard above this noise.

Listening at the Water Hole

The extra tweak that astronomers came up with in tuning their radio receivers was based on the realization that there is another interesting feature, as well as the radio noise from hydrogen, in this part of the electromagnetic spectrum. Many clouds of gas in space contain the so-called 'hydroxyl radical', written as •OH. It is in effect a molecule of water (H_2O) from which one hydrogen atom has been lost. Under the calm conditions in these clouds, it is stable enough to behave like a molecule in its own right. We know that •OH is there because we detect it by its characteristic radio emission at a frequency of 1,667 MHz. This is close to the frequency hydrogen broadcasts at, but far enough away to be distinctive. Hydrogen plus •OH adds up to water, H_2O, and the region between 1420 and 1667 MHz has been dubbed the electromagnetic 'water hole'. It is a region that would be bound to be interesting to any

A 200-metre diameter antenna operating at a wavelength of about 20–1 cm can detect signals from a similar antenna 1000 light years away.

creatures, like ourselves, for which water is important, and it is smack in the region of the electromagnetic spectrum where the Universe is dark and quiet. So, since about 1980, most searches for extraterrestrial life have concentrated on this part of the electromagnetic spectrum.

The hope is that, just as in semi-arid regions on Earth different species of animal meet when they come to drink at the water hole, so we may 'meet' aliens electronically by signalling (and listening for signals) at the wavelengths corresponding to the radio 'water hole'. There have been about 50 such SETI projects since Project Ozma, and none has found evidence of alien signals. But it is worth going into a little detail of how one of those projects was carried out, just to show how easy (and cheap) the whole thing is.

SETI in a Suitcase

It's worth bearing in mind that even the water hole alone offers millions of different frequencies to search, so it is no real surprise that we haven't struck lucky yet. The only real hope of getting a breakthrough is to automate the search, and that approach was pioneered by the American astronomer Paul Horowitz in the 1980s.

Horowitz designed a system, so compact that he called it 'Suitcase SETI', which could search across part of the band of frequencies in the water hole. One special feature of this search was to look for very precisely tuned bursts of radio noise, at very precise wavelengths (and frequencies), because he

THE INTERNET SETI PROJECT

Apart from having a radio telescope with which to detect faint signals from space, the big problem with SETI searches is to find the computer time to analyse all the data and pick out anything that might be an intelligent signal. An approach that caught the imagination of the public at the end of the 1990s was the SETI@home project, which hooked up more than a million home computers, via the Internet, to check data coming in from the Arecibo radio telescope.

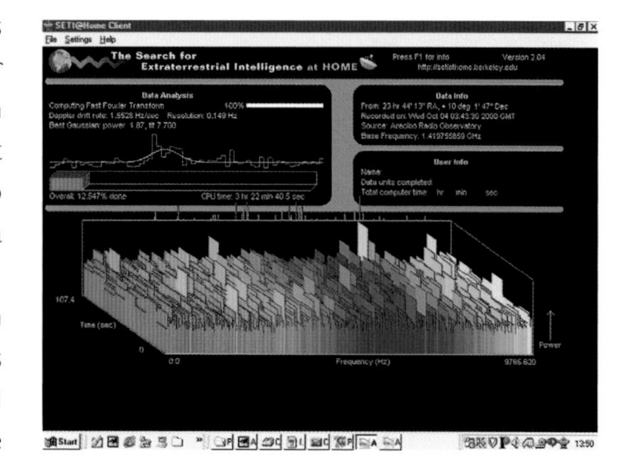

The Arecibo telescope scans a strip of the sky as the Earth rotates, from the equator up to a celestial latitude of about 35 degrees. Happily, this includes many of the star systems now known to have planets. The SETI equipment is linked up to the telescope and takes advantage of any free time, looking wherever the telescope happens to be pointed while the radio astronomers are using it for their regular work. Over several years, this means that it sweeps across the entire band of the sky covered by the telescope, pulling in a huge amount of data.

This is where the Internet comes in. The SETI@home team offered enthusiasts a screensaver (right) which they could download free into their home computers. This came with a small computer program which downloaded a tiny fraction of the huge mass of data from Arecibo, so that whenever the home computer was not in use it would quietly get on with analysing the data. More than a million people downloaded the software (from www.setiathome.ssl.berkeley.edu), which means that for at least part of the time, every day, over a million home computers are busily searching the Arecibo data for signs of intelligent signals from space.

1. The central region of our Milky Way galaxy gives an impression of just how many stars, and therefore potential homes for life, there are in our cosmic island.

reasoned that such precision could only be produced by advanced technology. The computer would search 65,000 radio channels, but each channel would be only 0.015 Hz wide. The project was funded by a private, non-profit organization called The Planetary Society. At first, the idea was to take the equipment from one radio astronomy observatory to another, fitting in with whatever research programmes were going on at the time. The best place to test the apparatus was, of course, the great dish of the Arecibo radio telescope, and the first run of Suitcase SETI listened out for intelligent signals from more than 200 of the nearest stars like our Sun.

After completing his tests (successfully, in that the equipment worked, but unsuccessfully as far as detecting extra-terrestrial life was concerned), Horowitz took his gear back to Harvard University where he worked. There, he discovered that an old 84-foot diameter radio telescope belonging to the university was about to be

The META project could detect a signal as weak as one ten-billionth of a watt of radio power arriving at the Earth.

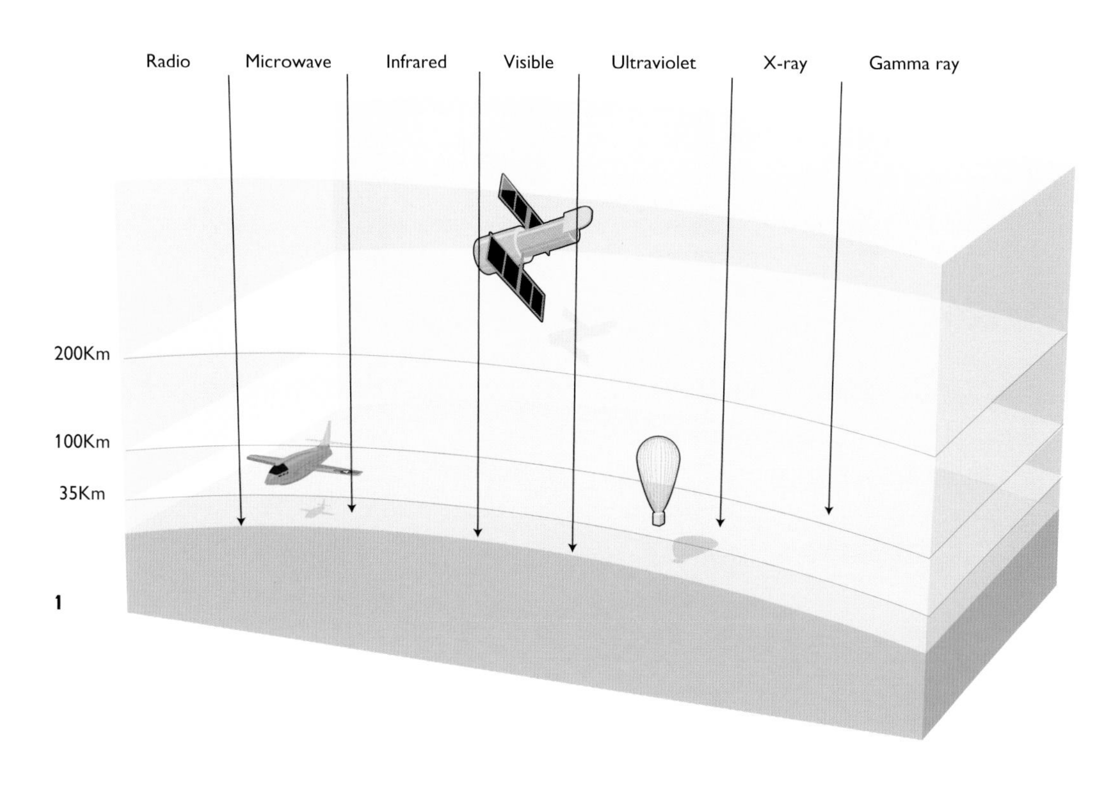

1. Different wavelengths of radiation penetrate to different depths in the atmosphere.

The most distant star yet studied directly for intelligent signals is less than 1 per cent of the distance across our Galaxy from us.

made obsolete. The Planetary Society agreed to fund the conversion of Suitcase SETI into a permanent facility, soon renamed Project Sentinel, using this telescope. It became operational in 1983, using a very simple technique to scan the heavens. Each day, the telescope's antenna would point at a chosen angle up to the sky while the Earth rotated. So it swept around a narrow strip of the heavens in 24 hours. Then the angle of the antenna was shifted a little (by half a degree) and the next strip of sky was searched. It covered 131,000 radio 'channels' (upgraded from the original 65,000) – impressive, but peanuts compared with the size of the water hole.

Spielberg's Contribution

The next improvement in the search for detecting extra-terrestrial life came courtesy of Steven Spielberg, director of the movie *ET*. He donated $100,000 to The Planetary Society and with this funding – absolutely tiny compared the cost of the space programme – Project Sentinel metamorphosed, in 1985, into Project META, which stands for Megachannel Extraterrestrial Assay. Technology had moved ahead so rapidly in the 1980s that META was able to search at 8,000,000 channels in the water hole, effectively resolving the problem of Doppler shifts once and for all.

There is still no sign of extra-terrestrial life, but there are still many millions of channels to search. And the standard answer to critics of the project is that it is cheap and funded entirely by voluntary donations. It is, therefore, potentially amazingly cost-effective. But this isn't the only way to search for life beyond the Earth.

BIG IS BEAUTIFUL

Projects such as Sentinel and META only look at a small patch of the sky at a time – the direction the antenna is pointing. The big worry about such an approach is that a signal might arrive from one direction in space while we are looking in another (although we hope that any intelligent aliens would keep broadcasting in the same place for months or years, giving us time to find the signal). One way to scan the whole sky (or interesting regions of the sky) quickly is to use more computer power to look in several directions at once. One way to increase the sensitivity of a detector is to use several different antennae linked together.

This 'brute force' approach was typified by a design study for Project Cyclops, carried out by NASA in the 1970s (the project itself never went ahead). The Cyclops proposal called for an array of 1000 radio antennae, each 100 metres in diameter, spaced 300 metres apart in the desert in the southwestern United States. With sufficient computer power, radio waves coming from different directions in space can be picked out separately by data-processing computers hooked up to the entire array. One processor could look in one direction while another looked in another direction at the same time.

The cost of such a system was clearly so high that there was never any prospect of it going ahead. On the other hand, it would have amounted to just three months of the expenditure the United States actually did pay out in the 1970s for the Vietnam war.

Into the Future

The big difference between the early twenty-first century and the days of the Project Cyclops design study is the huge increase in computing power available combined with the dramatic

THE FERMI PARADOX

If there are intelligent aliens in the Universe, with advanced technology, why aren't they here? This puzzle (it isn't really a paradox, in spite of its name) was put forward as an argument that we are alone in the Universe by the physicist Enrico Fermi (right) in 1950. The point is that the Universe and the Milky Way are billions of years old (at least twice as old as the Sun and Solar System) so there has been ample time for other intelligent civilizations to arise and to develop the means to travel across space. With computers and rocket technology only slightly more advanced than our own, such beings would be able to send probes, travelling at less than the speed of light, to visit every star in the Galaxy in a span of just 300 million years. They could do this because one probe would arrive at a system like our Solar System and use material from the moons and asteroids in it, to build many copies of itself to send on to explore other star systems. For the cost of just one probe (or two if you want a backup), you could explore the Milky Way.

The best answer to the puzzle is that, although the Universe is more than 10 billion years old, that is barely long enough for intelligence to have emerged. It took generations of stars to make the heavy elements that the Earth and its inhabitants consist of, and billions more years for an intelligent species to evolve. The counter to the Fermi Paradox is the argument that there ought to be many civilizations just about at our level, poised on the brink of space travel, but that we are unlikely to meet anyone much more advanced than ourselves. Nobody knows which point of view is correct.

The Arecibo radio telescope featured in the movie *Contact*.

decrease in its cost. This means that even without building such a large array we can still get many of the benefits of Cyclops. Although government funding is still not forthcoming for major SETI projects, they are now well within the means of private funding, as Project META demonstrated.

One system now being built is called, from its size, the One Hectare Telescope, or 1HT (1 hectare is 10,000 square metres). It will use a large number of linked antennae which can not only (thanks to that computer power) look at several dozen stars at the same time, but can also 'view' very many frequencies in the centimetre wave 'window' at the same time.

Even 1HT, though, is not an end in itself – it is a working testbed to prove the techniques that will be used to build the Square Kilometre Array (SKA), an array of telescopes in a grid 1km on each side (covering an area of a million square metres). SKA will be a hundred times more sensitive than any SETI detector yet built, which means that it can look ten times further out into the Galaxy and detect signals from a thousand times more star systems.

These projects are cheap because they use essentially the same data processing systems as home computers. The biggest problem facing the development of systems like SKA is that because technology has become so cheap and ubiquitous, the radio window is rapidly being filled up by the emissions from technology we are coming to regard as indispensable, such as mobile phones. It may be that by the middle of this century we will have to put our antennae on the far side of the Moon in order to find any signals from extraterrestrials without them being drowned by our own radio noise. Alternatively, cheap computers may become so powerful that they will be able to link together several systems such as SKA in different parts of the world and screen out human interference. Or maybe, just maybe, by 2050 that won't be necessary, because we will have already made contact.

1. An artist's impression of the I Hectare Telescope.

THE DRAKE EQUATION

In 1961, astronomer Frank Drake, of Cornell University, developed an equation (the Drake equation) which is a way of expressing the chance of there being other intelligent civilizations capable of communicating with us. It encapsulates a great deal of astronomy in a simple form, although people still argue about what numbers to put into the equation.

1. Some civilizations may destroy themselves before they get a chance to communicate.

2. Some aliens may have no wish to communicate. By some criteria, dolphins are intelligent but they have not developed technological civilization.

Counting Suns

The first parameter in Drake's equation is the rate at which new stars are forming in the Galaxy. This is one of the less contentious numbers and most astronomers are happy with a figure of about 20 per year, denoted by the symbol R.

How many of these stars will be similar to our Sun? Drake suggested that as many as one in ten may be sufficiently Sun-like for his purposes, and denotes this by the symbol f_s ('f' for 'fraction').

The next number in the equation assigns a probability that a Sun-like star will have planets (f_p). When Drake suggested a value (f_p) of 0.5, many people thought he was being remarkably optimistic, but recent discoveries of extrasolar planets suggest that he may have been right all along.

Counting Planets

The next item in the equation is pure guesswork: how many planets will there be in each planetary system that are suitable for life as we know it? The number of Earth-like planets (n_e) in our Solar System is exactly 1, of course, but some people argue that this is unusually large and others argue that it is unusually small. Just about the only guess we can make is to set the number at 1 and hope for the best.

What else do we need for technological civilization to arise on a planet? The first requirement is that there is life. The fraction of inhabitable planets which actually harbour life, denoted by f_l, is clearly less than 1, judging from our own Solar System and the barrenness of Mars. But it may not be *much* less than 1, judging from the evidence of complex organic molecules in interstellar clouds. Again, a guess is all we can come up with.

There are similar uncertainties surrounding the fraction of life-bearing planets on which intelligence arises (f_i) and the fraction of intelligent species that develop technological civilization (f_c).

This piece of the puzzle is particularly interesting. It is quite possible to have a civilization (such as the Ancient Romans, or the Incas) without having the kind of technology needed to communicate across interstellar space, and you can imagine a world in which civilization stays in that state indefinitely. It is also entirely possible to be intelligent without having technological

4

civilization – in some ways, dolphins are intelligent, but they don't have technology.

Counting Earths

The final piece of the puzzle concerns the longevity of such a civilization – its lifetime, *L*, expressed in years. In the 1960s and 1970s, the threat of nuclear war made some pessimists set this figure very low – no more than 100 years. Today, many people are more optimistic, but this remains a big uncertainty. Indeed, in the eyes of the pessimists we have simply replaced the threat of nuclear annihilation by the threat that we will destroy our planet through pollution, bringing an end to our present civilization before the end of the twenty-first century.

When you put everything together, you have an equation for the number of advanced civilizations actively involved in communicating across the Galaxy today.

$$N = R f_s f_p n_e f_i f_i f_c L$$

This is Drake's equation. If *R* is about 20, and the next six factors multiply up to give a value of about 0.05, which is plausible, what you end up with is the conclusion that *N* is roughly equal to *L*. This means that if civilizations do avoid blowing themselves to bits or polluting their planets to destruction, there may be a very large number of them out there trying to make contact with one another, and with us.

3. Even without warfare, technological civilizations may find ways to destroy themselves and the wildlife around them.

4. Intelligent life could exist deep beneath the clouds of a giant planet like Jupiter without even knowing that the rest of the Universe exists.

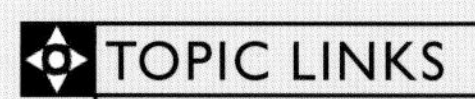

3.1 Life and the Universe
p. 139 Other Worlds
p. 148 Other Earths?
p. 150 Seeking Signs of Life

3.3 Listening to the Stars
p. 172 Life As We Don't Know It

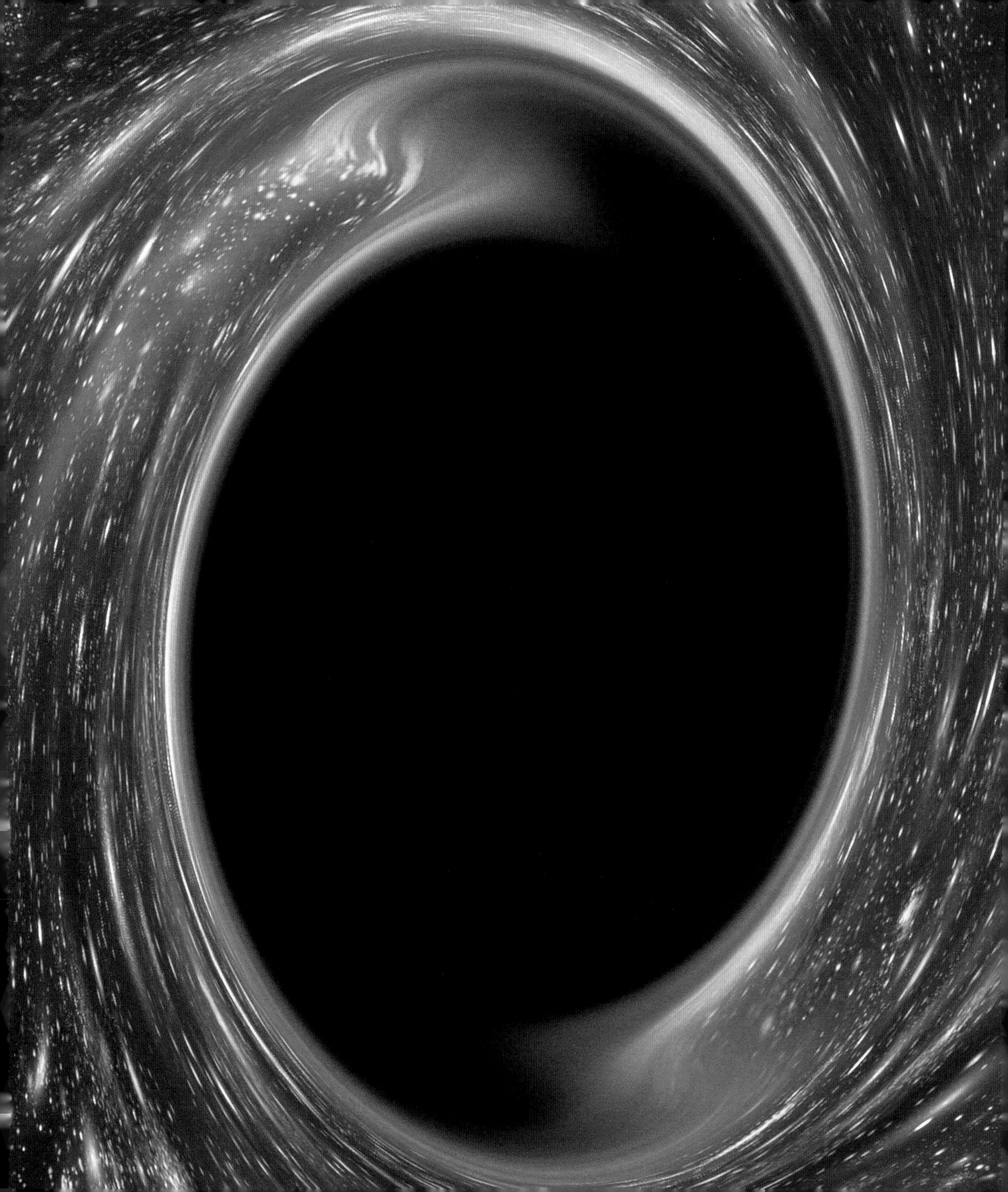

4

OTHER WORLDS

4.1 COSMIC COINCIDENCES

In many ways, the Universe we live in seems to be well suited to the existence of life forms like ourselves. To some extent this must be because we have evolved and adapted to suit our surroundings. But at another level, it turns out that life can only exist because of delicate balances in the fundamental laws of physics. Is this just a coincidence, or is there some deep connection between the way the Universe works and the presence of life in the Universe? Some people think that the answers to these questions lie outside the realm of science, but scientists are forever pushing back the frontiers of their domain. Although these questions are at the cutting edge of cosmological thinking today, and there are no definite answers, scientific exploration of such questions has already begun and science has now answered many questions previously thought to lie in the realm of philosophy.

Previous page. Computer simulation of the appearance of a black hole against a background of stars.

THE GOLDILOCKS UNIVERSE

The Earth is often referred to as the 'Goldilocks planet' because, just as baby bear's porridge was just right for Goldilocks, so the Earth is just right for life. But some astronomers go further than this, arguing that the existence of a planet like the Earth depends on such a precise set of conditions at large that we should think in terms of a 'Goldilocks Universe', set up (by accident or design) to be just right for life. The exploration of the implications of this idea goes by the name 'anthropic cosmology'. It is an exploration which is almost entirely in the mind, since there are very few ways in which the principles of anthropic cosmology can be put to the test.

1

2

1, 2 and 3. The Earth seems to be an ideal home for many varieties of life.

Life in the Universe

At its simplest, anthropic cosmology is simply a way of inferring what the Universe at large must be like from the fact that we exist. The best example (although it was not thought of in these terms at the time) is the way that Fred Hoyle predicted the existence of the resonance in carbon that allows the triple-alpha process to convert helium into carbon inside stars (▷ p. 54). In effect, Hoyle reasoned, 'We exist, and we are made of carbon compounds. So carbon must be manufactured in stars. So there must be a resonance that allows the triple-alpha process to proceed.' Nobody had ever measured this resonance or predicted its existence until Hoyle came along, and his prediction was based entirely on anthropic reasoning.

In 1957, just after Hoyle's prediction had been triumphantly confirmed by experiments, Robert Dicke pointed out that even the size of the Universe is 'not random, but conditioned by biological factors.'

3

Weak anthropic principle: The observed values of all physical and cosmological quantities take on values restricted by the requirement that there are sites where carbon-based life can evolve.

1. Life on Earth is well represented by a green, lush, moss-covered rainforest.

Dicke argued that, in order for us to exist, the minimum requirement is one star, with one planet, made of a suitable mixture of chemicals. The rest of the Universe, hundreds of billions of galaxies containing hundreds of billions of stars, spread across thousands of billions of light years, seems unnecessary. But remember that only hydrogen and helium (plus dark matter) emerged from the Big Bang (▷ p. 131). It took many billions of years for the first stars to manufacture the heavy elements, explode and scatter them across space, and many billions more years for life to evolve, on a planet created out of that stardust, to the point where it could ask questions about the Universe. All the while, the Universe was expanding. The fact that we are able to ask such questions implies that the Universe is many billions of years old, and many billions of parsecs across.

The Weak Anthropic Principle

Interest in anthropic cosmology only developed in the 1970s, when Brandon Carter elaborated on these ideas and divided them into two categories. The ideas of Hoyle and Dicke fit within the framework of what Carter called the 'weak' anthropic principle. This is based on the idea that the Universe we see around us is not the only universe that could exist. The mathematical models of the cosmologists tell us that it is possible to describe different kinds of universe, where there are different laws of physics (▷ p. 196). In one such model, for example, gravity might be stronger or weaker than it is in our Universe; in another there might be no resonance that allows the triple-alpha process to do its work.

PALEY'S WATCH

The eighteenth-century philosopher William Paley argued that the existence of living things so beautifully fitted to their environment as a human being or a flower required the existence of a designer; just as a man who knew nothing about watches (or marine chronometers, right) but found one lying on the ground could infer from the way everything fitted together inside it to make a working machine that it had been designed by an intelligent being. Just taking a heap of watch components and throwing them together at random, the argument runs, could never produce a working watch. The theory of evolution by natural selection removes the force from Paley's argument, because natural selection acts as the (unintelligent) designer that fits living things so beautifully into their ecological niches.

The important point is that, although evolution is unintelligent, it is not random. Individual changes in living things (mutations) are random, but evolution selects those changes that give an individual an advantage. Given a long enough span of time, this could convert a bacterium into a human being. Paley didn't know about evolution by natural selection, because he died in 1805, 53 years before Charles Darwin and Alfred Wallace published their theory.

A few cosmologists try to use the same argument to say that the Universe it must have been made by a designer. Most, however, think that we are in the same position as William Paley: as yet ignorant of the scientific rules which make the Universe look like a put-up job, even though it has evolved naturally. Hopefully, it won't be another 53 years before we find out what is really going on.

Strong anthropic principle: The Universe *must* have those properties which allow life to develop within it at some stage in its history.

Where could these other worlds be? If the Universe is infinitely large, there could be regions in which these different laws of physics operate, beyond the range of our telescopes (*forever* beyond the range of our telescopes, in regions of space that are being carried away from us faster than light by the expansion of the Universe). Or, if there have been many cycles in a bouncing Universe, there might have been different laws of physics 'before' the Big Bang.

Whatever possibility you can imagine, the weak anthropic principle says that there are lots of different universes (perhaps infinitely many) separated in space or time (or both) and that life can only exist to notice what is going on in universes very similar to our own.

But there is an alternative, which Carter called the 'strong' anthropic principle. This says that the Universe that emerged from the Big Bang, our Universe, really is unique, and that it had no 'choice' about the laws of physics; they had to be just right for the presence of life – specifically, for human life.

These lead some cosmologists into the strange world of quantum mechanics, where, according to some interpretations of what the equations mean, reality does not exist unless it is observed by intelligent beings. In this (weird) argument, the laws of physics *have* to be what they are, so that we can exist, so that we can notice what the laws of physics are, and make them real. Others see the existence of the 'coincidences' that make the Universe just right for life as evidence that it was designed. This still leaves the question of where the designer(s) came from.

The strong anthropic principle leads us into the realms of philosophy and religion and away from science. Fortunately, however, the weak anthropic principle is all we need to see how cosmological investigations are likely to proceed over the next few years.

A Tailor-made Universe?

The extent to which our Universe is 'just right' for life can best be seen by looking at how different it might be if small changes were made in one or more of a number of key parameters which shape the Universe. Physicists discuss the influence of 20 or more of these parameters, some of them quite exotic by everyday standards. But there are just a handful of more obvious candidates which make the point. None of them does so more forcefully than the force which literally shapes the Universe – gravity itself.

The Role of Gravity

Gravity seems so important to us, and is so important to the Universe at large, because it is always additive. Every atom and subatomic particle in the Earth contributes its tiny influence to the overall gravitational pull of the planet. The same is not true of the other familiar everyday force, electromagnetism. Atoms contain both positively charged protons and negatively charged electrons, so overall both the atoms and the Earth have zero electric charge. But gravity is actually an incredibly feeble force, which can best be seen by comparing the gravitational force between two protons with the electric force between the same two protons.

Because both electricity and gravity obey inverse square laws, the ratio of the two forces is always the same, no matter how far apart (or how close) the two protons are. The electric force that keeps them apart is 10^{36} times bigger than the gravitational force. This large number is so important that the physicists who ponder its implications simply refer to it as N. This number alone explains why stars are so big.

1

Imagine starting out with a set of objects containing, respectively, 10 hydrogen atoms, 100 atoms, 1000 atoms and so on, packed together. The 24th object would contain 10^{24} atoms and be about the size of a sugar cube. The electric forces that shape atoms would have no trouble resisting the inward tug of gravity, which has a 1036 'handicap' over electricity. The 39th object would be a kilometre across, and still able to hold itself up against gravity.

But the volume and mass of material goes up as the cube of its radius (essentially, as the cube of the number of atoms), increasing the gravity at the centre by three powers of ten, while the inward tug of gravity goes down as the square of radius (two powers of ten) as the object gets bigger. Overall, gravity gains on electricity as a two-thirds power. By the time we reach the ▷▷

1. Even a small amount of electricity can make your hair stand on end, resisting the gravitational pull of the whole Earth.

Overleaf. Multiple bolts of lightning pictured over Tucson, Arizona.

54th object, which would be about as big as Jupiter, gravity will begin to crush atoms into their component parts at the centre, because 54 is three-halves of 36 (▷ p. 31). The 57th object in the collection will be so big that gravity in its heart is enough to make protons bash together hard enough for fusion to occur. Stars like the Sun do indeed contain about 10^{57} protons.

The Big ε

The strength of gravity determines how big stars are. Another fundamental number determines how long they live. When four protons (hydrogen nuclei) are converted into one nucleus of helium-4, 0.7 per cent of the mass of the protons is released as heat (▷ p. 53). This proportion, 0.007, is a measure of the strength of the strong force that holds nuclei together in spite of the electrical repulsion between protons. All the rest of the nuclear fusion process inside stars, from helium up to iron, releases only a further one-seventh as much energy as the initial fusion of hydrogen into helium, so the lifetime of a star depends almost entirely on the number 0.007 (often written as ε), which determines how much energy is available to keep the star shining.

But E does more than that. If the number were just 0.006, the strong force would be too weak to hold nuclei of deuterium (a proton plus a neutron) together. The first step in the fusion process (the pp chain ▷ pp. 58–9) would not occur, and there would be nothing more complicated than hydrogen in the Universe. Alternatively, if E were just 0.008, the strong force would be so strong that two protons could stick together. Further fusion

FROM ONE EXTREME TO ANOTHER

Not everybody likes the idea of anthropic cosmology. In his book *Perfect Symmetry*, published in 1985, the eminent physicist Heinz Pagels (right) wrote: 'Physicists and cosmologists who appeal to anthropic reasoning seem to me to be gratuitously abandoning the successful program of conventional physical science of understanding the quantitative properties of our universe on the basis of universal physical laws. Perhaps their exasperation and frustration...has gotten the better of them...The influence of the anthropic principle on the development of contemporary cosmological models has been sterile. It has explained nothing, and it has even had a negative influence, as evidenced by the fact that the values of certain constants, such as the ratio of photons to nuclear particles, for which anthropic reasoning was once invoked as an explanation can now be explained by new physical laws...I would opt for rejecting the anthropic principle as needless clutter in the conceptual repertoire of science.'

At the other extreme, typifying the kind of argument that annoyed Pagels, Fred Hoyle sees the Universe as 'a put-up job', and wrote in his book *Galaxies, Nuclei and Quasars*, published in 1965, that: 'The laws of physics have been deliberately designed with regard to the consequences they produce inside stars. We exist only in portions of the universe where the energy levels in carbon and oxygen nuclei happen to be correctly placed.'

Both viewpoints may be too extreme. Perhaps we live in a Universe where the energy levels, and other parameters, are indeed 'just right', but by accident rather than design. That would leave scope for Pagels wish 'that fundamental laws which determine the nature of the Universe can be discovered by the traditional methods, but remove the need for a designer.'

1. Is the Universe made for humans, or are humans made for the Universe?

could occur, but there would be no hydrogen in the Universe (only helium-2), and therefore no water – an essential pre-requisite for life as we know it. This is one of the most powerful examples of the Goldilocks effect at work.

The Question of Q

Another way of seeing the special nature of the Universe is to look at its roughness, which is measured by a number the astronomers refer to as Q. If the Universe had emerged perfectly smoothly from the Big Bang, then it would still be perfectly smooth – a uniform sea of gas expanding in all directions.

In fact, when astronomers measure how much energy would be required to break up systems like clusters of galaxies and disperse them, and compare this with the total mass energy ($E = mc^2$) in the same system, they find that the two numbers are in the same ratio – 1:100,000 – right across the Universe. In other words, Q, the roughness of the Universe, is 2^{-5}. When they measure the size of the ripples in the cosmic microwave background radiation, astronomers find the same number. The roughness of the Universe has been 1 in 100,000 ever since the beginning of time.

If Q were much bigger than this, gravity would have pulled clumps of matter together very early on in the life of the Universe, forming supermassive stars and black holes. ▷▷

THE 3-DIMENSIONAL UNIVERSE

Surprising though it seems when you first encounter the idea, one of the most special things about our Universe that helps to make it suitable for life is that it exists in three dimensions of space. Although time is regarded as a fourth dimension, it behaves in a different way to the three space dimensions, and it turns out that if there were either more or less than three dimensions of space life would be impossible.

2

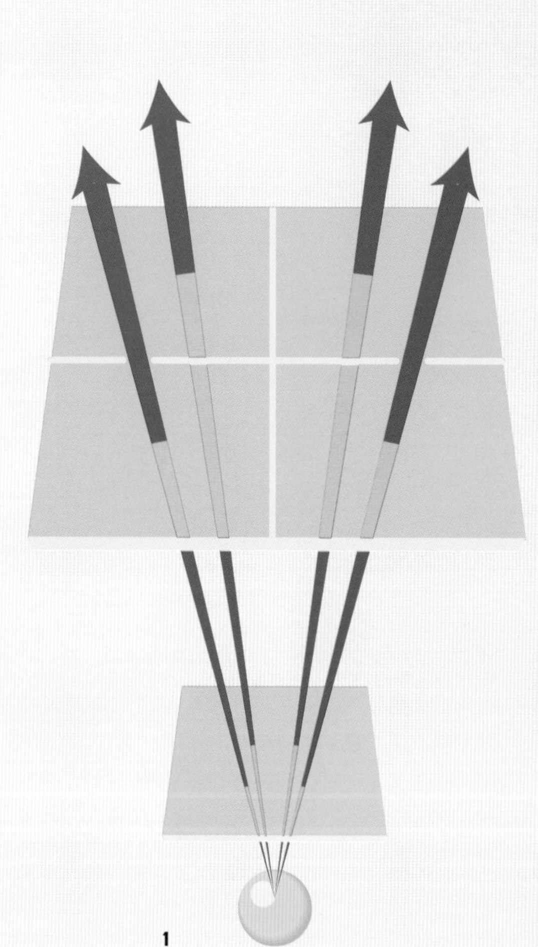

1

Feel the Force

In our three-dimensional Universe, forces such as gravity and electromagnetism obey inverse-square laws. The electrical pioneer Michael Faraday had a nice way of explaining this.

Imagine 'lines of force' reaching out evenly in all directions from a charged particle. If you envisage a sphere surrounding the particle, a certain number of lines will cross each square centimetre of the surface of the sphere. If the radius of the sphere is bigger, its area increases as the radius squared. So we still have the same number of lines of force, the number of lines crossing each square centimetre goes *down* as the radius (the distance from the centre) squared. Electricity obeys an inverse-square law.

Four is Too Many

If there were four spatial dimensions, the area of a 'four-sphere' would go up as the cube of its radius. So electricity (and gravity) would obey inverse *cube* laws. Double the distance and the force would go down by a factor of eight (2^3) not four (2^2).

This is bad news, because an inverse-cube law (or any law apart from inverse square) does not allow stable orbits for planets. In our Universe, a planet that receives a slight nudge stays more or less in the same orbit, because an inverse-square law exactly balances out the centrifugal force and gravity and the centrifugal force stay in balance over a comfortable range of distances. In a four-dimensional universe, the force changes more rapidly over a short distance (it is said to have a 'steeper gradient'), so that just inside any particular orbit it overwhelms the centrifugal force, while just outside the orbit the centrifugal effect dominates. So a planet that received even the tiniest nudge would either plunge into its sun, or spiral away into the depths of space. And in higher dimensions still, the effect is even stronger.

Two is Not Enough

In a two-dimensional universe, complex life could not exist at all. You might imagine a kind of flat creature, like an amoeba, that could eat by opening a hole in its side and

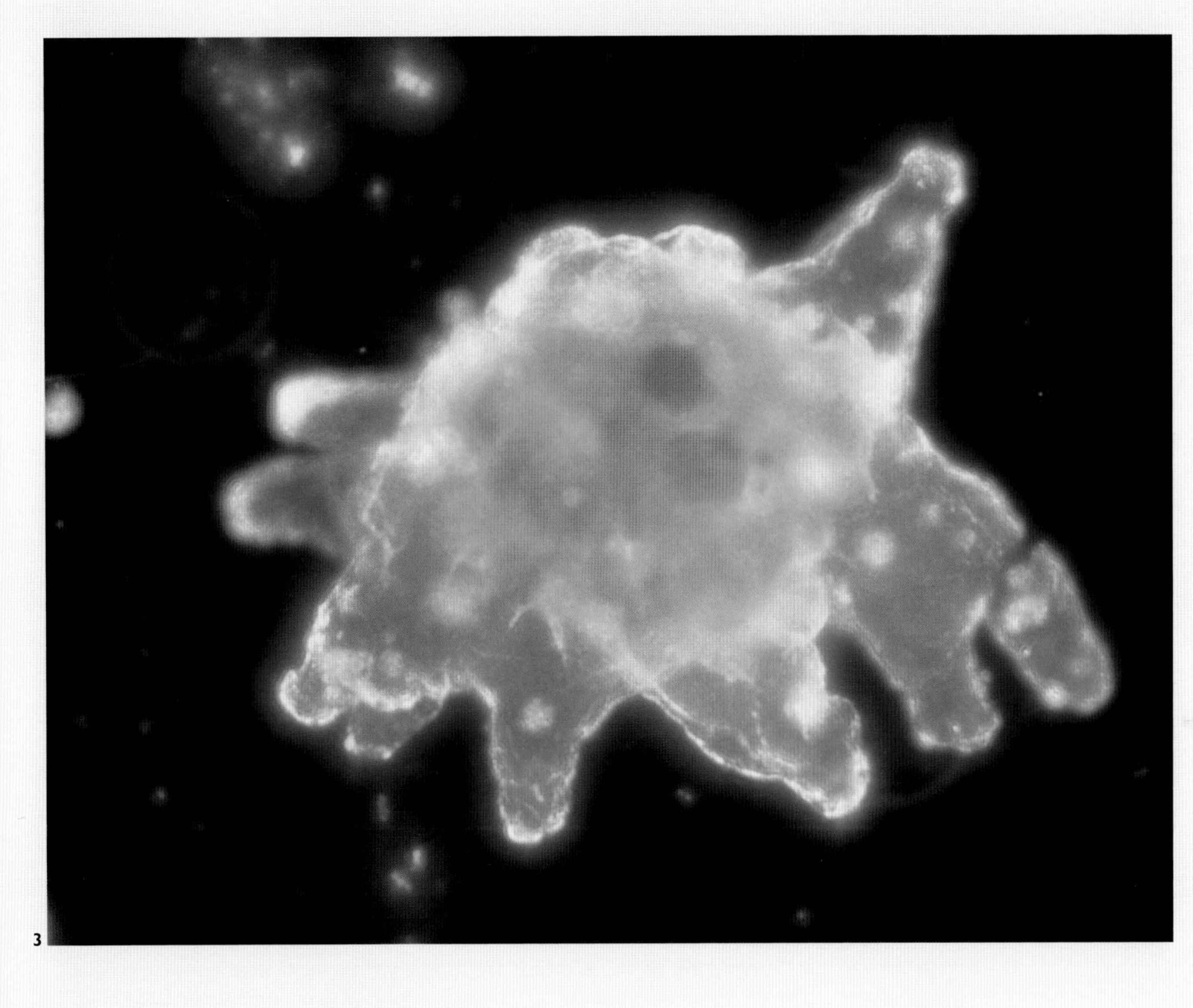

3

1. One way of visualising the inverse square law. If you double the distance, lines of force are spread over four times (2^2) the area.

2. Michael Faraday, who originated the idea of line of force.

3. An amoeba is the nearest thing on Earth to a two-dimensional life form.

swallowing things, then opening another hole to excrete (it couldn't open both holes at once, or it would fall apart). But it could only have a very simple brain, because lines could not cross one another, so there could be no connections between neurons except in the simplest way (and even simple connections would block the flow of food).

The number of space dimensions has to be *precisely* three if intelligent life is to exist in the Goldilocks Universe.

TOPIC LINKS

4.1 Cosmic Coincidences
p. 187 Paley's Watch
p. 188 The Role of Gravity

4.2 A Choice of Universes
p. 211 Geography and Relativity

☆ There are roughly a billion (10^9) photons (particles of light) in the cosmic background radiation for every single baryon (proton or neutron) in the Universe.

It would have been a very different place from the Universe we live in. The fact that Q is just big enough to allow interesting things to form in the Universe, but that the Universe is very nearly flat (▷ p. 95), is part of the Goldilocks effect. It is also intimately connected with the way the Universe was driven outwards by inflation very early on in its life (▷ p. 115).

WHAT MIGHT HAVE BEEN

Because gravity is such a feeble force, compared with the other forces of nature, you might think that the exact strength of gravity doesn't matter very much. But you'd be wrong. The mass of a star is determined by the balance between gravity and electrical forces, and depends on the square root of the cube of the strength of gravity (a so-called 3/2 power law), because the volume and mass of the star go as the cube, and gravity is an inverse square law. This means that if the strength of gravity were increased by a factor of a million (10^6), the mass of hydrogen needed to make a star would be reduced by a factor of a billion(10^9), because 9 is 3/2 of 6.

Such a change produces a dramatic effect on the kind of conditions that exist in a cosmologist's model universe. Look at what might happen in a universe where the strength of gravity is a million times stronger than it is in our Universe, but all the other key parameters stayed the same.

High-speed Stars

In our Universe, the Sun is a typical star, with a mass defined as one unit (one solar mass) and a lifetime of about 10 billion years (1010 years). In our alternative universe less material is required to produce the same pressure and stars will get hotter at lower mass, so stars will typically have masses about one-billionth (10^{-9}) of the mass of the Sun, which corresponds to about 10^{18} tonnes, a bit less than 1.4 per cent of the mass of our Moon. Everything else going on inside the star (including the nuclear reactions that release the energy to hold it up) is just the same as inside the Sun, and an atomic nucleus at the heart of the star doesn't 'know' much about the strength of the gravitational force in absolute terms, only about the weight of material pressing down on it.

The lifetime of a star is related to its size, not simply to the amount of fuel present, because, by swelling or shrinking slightly as appropriate, it adjusts the rate at which fuel is burnt to compensate for the energy being lost from its surface. So what really matters is how long it takes electromagnetic energy to get from the centre of the star, where it is generated, to the surface, where it escapes. Because the radiation is bounced around inside the star on its way to the surface (rather like the way a ball gets bounced around in a complicated pinball machine), this depends on the square of the radius of the star, not on the radius itself. Mass goes up as radius cubed, so stars with masses 10^{-9} times the mass of the Sun will have radii 10^{-3} times the radius of the Sun, and lifetimes only 10^{-6} times the lifetime of our Sun – about 10 thousand years, rather than 10 billion years.

Could Life Exist?

With all other processes, and in particular chemistry (the way atoms and molecules interact, unchanged in our model universe, it is hard to see how life would have much chance to evolve on a planet orbiting such a star before the star exhausted its nuclear fuel and went through the usual stages of swelling up to become a red giant then fading away to become a white dwarf. Just for fun, however, imagine what life would be like if it did exist on a planet with the same surface gravity as the Earth in such a universe.

1

1. Inside a star, protons bounce around like balls in a souped-up pinball machine.

1. (opposite) Could our Universe be just one bubble in a multitude of universes each with its own set of physical laws?

The same sizing factor applies, of course, to planets as well as to stars. In order for the gravity at the surface of a planet in the model universe to be the same as it is on Earth, the planet would have to have a billion times less mass than the Earth. It would weigh in at about 5000 billion tonnes (5 x 1012 tonnes) – about the mass of a typical asteroid in our Universe. But just because the pull of gravity at the surface of such an object would be the same as at the surface of the Earth doesn't mean you could walk about there. The mass of your own body also has to be taken into account when working out the force that holds you down on a planet (your weight), and the gravitational pull of your own body would also be hugely increased in the imaginary world. The extra strength of gravity would be enough to crush any human-sized object out of existence. There would be no great mountains or tall trees on such a planet. Any animal life forms that did exist would be low and squat, close to the ground and walking carefully on several stumpy legs to support their own weight. Falling over, even from a height of a few centimetres, would be a disaster.

A Compact Universe

As if that weren't bad enough, it is extremely unlikely that planets could exist in stable orbits in this compact universe we are imagining. Because of the greater strength of gravity, galaxies would form very early on in such a universe, from clumps of gas physically smaller than the clumps which formed galaxies in our Universe, but with the same gravitational attraction. Stars would be much closer together within those galaxies. There would be roughly the same number of galaxies as in our Universe (at least a hundred billion) but each one would be smaller in diameter than the Milky Way by a factor of about a million. Instead of being about 30,000 parsecs across, a typical galaxy would be about 0.03 parsecs across (about 10^{12} km in diameter, less than the distance from the Sun to Alpha Centauri). Instead of stars being separated from one another by distances of parsecs, they would be so closely packed that close encounters between them would be common. The gravitational influence of passing stars would tug any planets that did form out of their orbits and scatter them through interstellar space.

When would all this be happening? Our own Universe is at the stage where the heavy elements have been produced in the first generation of stars and have been converted into interesting things like ourselves. It has also been expanding for some 14 billion years. In that sense, it is 14 billion light years from here to the most distant thing we can see (a number sometimes known as the Hubble radius). In the compact universe, the same interesting stage of heavy element production and distribution (although probably not, alas, interesting things like life) will be reached a million times quicker, after just 14,000 years, when the universe has a Hubble radius of 14,000 light years (a diameter of 28,000 light years). In other words, the high-speed, compact universe would reach the stage our own Universe is at today when it is less than a third of the size of the Milky Way.

There is nothing in the laws of physics to say that such universes, and others even more bizarre, cannot exist. So where are they?

Overleaf. Artist's impression of our Milky Way Galaxy.

4.2 A CHOICE OF UNIVERSES

If our Universe has not been specifically designed for us (a 'bespoke' Universe), the fact that it is just right for life may be because it is just one universe among many. In the same way, the fact that a suit fits may not be because it is tailor-made just for you but because you have chosen it from a variety of ready-made suits. For this idea to work, our Universe cannot be unique. There must be a huge variety of universes, somewhere in space and time, like the variety of different suits hanging on the racks in a shop. There may be uncountable sterile universes (suits that do not fit), but life can only exist in universes like our own (suits that do fit). To explore this possibility, cosmologists have to get to grips with the strange world of quantum physics. Quantum physics naturally includes the idea of many worlds, existing in some sense side by side. This has long been a source of delight to science fiction writers and their readers, but now these non-commercial ideas have to be considered within the frameworks of science fact.

THE MANY WORLDS OF THE QUANTUM

Quantum physics is strange to us because it describes what goes on on tiny scales, smaller than atoms, and we have no everyday experience of what things are like inside atoms. What we do have are equations that describe what goes on – the equations of quantum mechanics. These equations are completely reliable if they are used to predict how measurable things (like the positions of the electrons used to paint a picture on your TV screen) change. Unfortunately, there are several (at least half a dozen) so-called 'interpretations' of quantum mechanics, which try to explain what is happening to things like electrons when they are not being measured. All are equally valid as aids to human imagination, though none is 'the' truth. The interpretation many cosmologists prefer, for reasons that will become clear, is called the 'Many Worlds Interpretation'.

Particles and Waves

Quantum weirdness is clearly demonstrated by the way things like electrons and photons behave both as particles and as waves. Experiments that are designed to measure the waviness of quantum entities (anything smaller than an atom) clearly show that they are waves. The classic example is the 'two slit' experiment, where a beam of light is shone on a screen with two holes in it. Light spreads out as waves from each of the holes

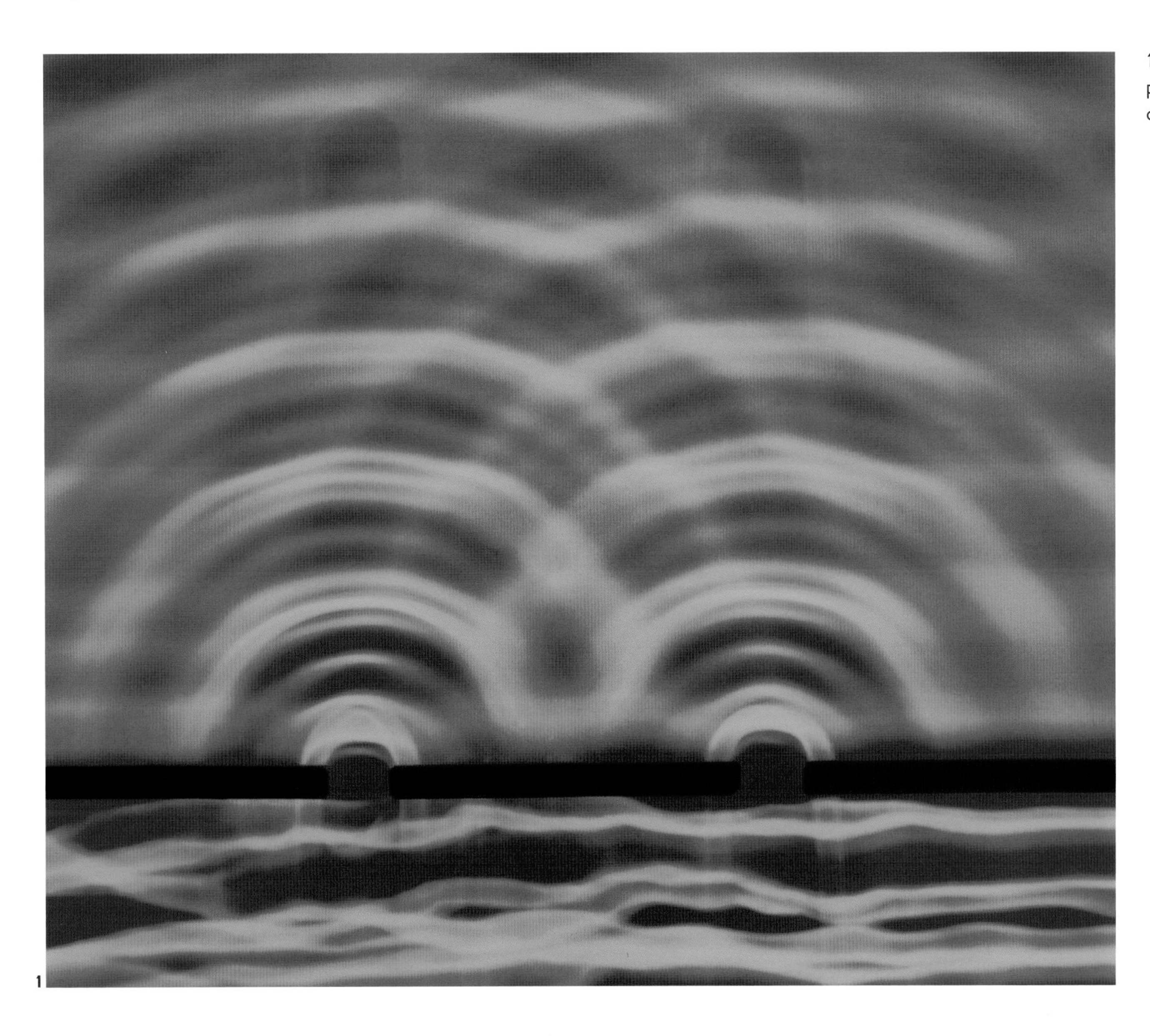

1. Interference pattern produced by two sets of waves.

and forms a characteristic pattern of light and shade (an 'interference pattern') on a second screen. This exactly echoes the way ripples on a pond interfere with one another. Similar experiments with beams of electrons show them behaving as waves.

But when experiments are designed to measure the particle properties of electrons or photons, recording their arrival at a target like a stream of little bullets, they turn out to be behaving just like a stream of little bullets. Quantum entities are both particle and wave – a phenomenon called 'wave–particle duality'. It seems that quantum entities travel like waves, but arrive as particles. The equation that describes how they move is called a 'wave function'.

Many Worlds

Wave–particle duality is only the beginning of the story of quantum mysteries. The way a travelling quantum entity 'decides' what kind of particle it is when it is observed is also unlike anything in our everyday experience. As a simple example, think of a single electron travelling through space. When electrons are measured, among other things they each have a property called 'spin'. This is not like the spin of a top, or the Earth on its axis. It can best be thought of as a 'label' on the electron. All that matters is that the measured spin of an atom can only have one of two values, called 'up' and 'down'. A measured electron always has one, and only one, of these spins.

What of a travelling electron? The standard interpretation of quantum theory says that when an electron is on its own it does not have a definite spin, but exists in an indeterminate 50:50 state, a mixture of spin up and spin

1. (opposite) A rainbow is produced when lightwaves are refracted and reflected by raindrops.

THE REALITY OF DUALITY

People sometimes think that wave–particle duality is simply a statistical effect. After all, the waves on the ocean are actually made up of billions of tiny particles – atoms. But quantum wave–particle duality is not like that – it operates at the level of individual entities such as photons and electrons, with the 'waves' having about the same size as the 'particles'. Although there had been indirect evidence of this phenomenon since the 1920s, the reality of duality was brilliantly confirmed in the early 1990s, in an experiment devised by a team of Indian theorists and carried out by a team of Japanese experimental physicists. It took so long to provide this final confirmation, because until then the technology was not up to the task. The experiment involved sending single photons (individual particles of light) through a tiny air gap between two blocks of glass (two prisms), and monitoring their behaviour.

The experiment required great precision, not just in producing single photons but targeting them through the gap between the two prisms, with the size of the gap controlled to within a few tens of a billionth of a metre, about one-tenth of the wavelength of the light involved. Because the gap is so small and light travels as a wave, it could cross the gap, but other tests showed that the photons that arrived on the other side of the gap were particles. Individual photons had been observed behaving as both wave and as particle in the same experiment. Dipankar Home, the leader of the Indian team, summed up the implications. 'Three centuries after Newton,' (left) he said, 'we have to admit that we still cannot answer the question "what is light?"'

1. John Wheeler, who encouraged Hugh Everett's work on the 'Many Worlds' idea.

2. According to quantum theory, the Universe is like a many-branched tree, with each branch corresponding to a different reality.

down. This 'superposition of states', as it is called, applies to all quantum properties. It is only when a measurement is made that there is a 'collapse of the wave function' and the electron chooses (at random) which spin (or whatever) to have. Einstein hated the idea of this random behaviour, and famously declared 'I cannot believe that God plays dice.'

But there is an alternative interpretation. Instead of one electron in a superposition of states, the equations work just as well if there are two electrons, one in each possible state, but in two separate universes. Whichever universe you are in, if you measure the spin you get a definite answer, with (in this case) a 50:50 probability of either choice. For more complicated situations (for example, throwing six dice at once), the odds are more complicated, but there is no collapse of the wave function. On this picture, as Dorothy Parker might have said, 'an electron is an electron is an electron.' The cost is that there is now a separate universe – a separate physical reality – for every possible outcome of every possible quantum measurement. This is where the name 'Many Worlds Interpretation' comes from.

Quantum Physics that Suits Cosmologists

Cosmologists like the idea of many worlds because they have great difficulty with the standard interpretation of quantum mechanics. It is possible to imagine the whole Universe being described mathematically by a single quantum wave function. Such a wave function is sometimes known as the Wheeler–DeWitt equation, after two physicists who studied the problem, but we don't actually know what the equation is, only some of the properties it must have.

The problem cosmologists then have with the standard interpretation is that the Universe is everything there is, so there is nothing outside the Universe to interact with the wave function and make it collapse – it has to exist forever in a superposition of all possible states. This is essentially the same as saying that all possible universes exist side by side, and all instants of time exist (one in front of the other), with nothing really changing. This literally means that the science fiction idea of alternative histories (a world where the South won the American civil war, a world where Nelson Mandela was never imprisoned, and so on) becomes part (a rather small part) of the quantum mechanical description of reality. What are the implications for ideas such as the Big Bang and the expanding Universe?

Quantum Cosmology

There are two kinds of quantum cosmology. The first kind deals with events that occurred very close to the Planck time (time zero) in 'our' Universe (▷ p. 98), and which led to inflation and the Big Bang in which hydrogen and helium were created out of pure energy in a little less than four minutes (▷ p. 107). The second kind of quantum cosmology deals with the implications of the Many Worlds Interpretation and the Wheeler–DeWitt equation. It is much more speculative and exploring these ideas takes us beyond the frontiers of what is known about the way the world works. But it is this kind of exploration that makes the unknown known, and which has already made the first kind of quantum cosmology a respectable part of astronomical thinking.

★ After his PhD, Hugh Everett worked on classified subjects for the Pentagon and never published another scientific paper.

2

1. Stephen Hawking, a pioneering quantum cosmologist.

2. To a polar bear standing at the North Pole, all directions are south.

1

In 1906, J. J. Thomson received the Nobel Prize for proving that electrons are particles. In 1937, his son, George Thomson, received the Nobel Prize for proving that electrons are waves. Both were right.

The Cosmic Desert

Accepting, for the moment, the everyday idea of time flowing inexorably forward, we can picture the cosmic version of the Many Worlds Interpretation as implying that the Universe was split into many different branches, like a huge tree (sometimes called the 'Multiverse'), by quantum processes operating at the beginning of time, when what is now the observable Universe was the size of a quantum entity. The different branches of the Multiverse would still be in some sense members of the same family, and governed by some common overriding principles (especially the quantum principles which describe the splitting process and allow the universes to keep branching repeatedly as time passes). But there will be an enormous number of different universes (possibly infinitely many) within the Multiverse, and among that infinite array there will be universes with all possible values (and combinations of values) of the fundamental parameters such as omega and lambda, or the Hubble parameter. There could even be universes with different values of N, or E, or Q.

In the vast variety of all the possible branches of the Multiverse, the combinations of the fundamental parameters will allow life to exist in only a tiny proportion of the universes. (If the Multiverse is infinite, life might still exist in an infinite sub-set of universes, but that infinity would only be a tiny proportion of the bigger infinity.) This is where the weak anthropic principle (▷ p. 187) comes into its own. Most of the Multiverse is sterile – a cosmic desert. Life exists only in a relatively small number of oases scattered across the cosmic desert. But, since we are living things, when we look around us we have to see an oasis, not the desert. These ideas can even predict what kind of oasis we ought to see.

Hawking's Universe

Stephen Hawking developed a version of cosmology based on the Many Worlds idea in the early 1980s. His starting point was the assumption (or guess) that there must be no boundary to the Universe – no 'edge' either in time or space. From the conventional point of view, as we have seen, there is an edge of time at the Big Bang (or, strictly speaking, at the Planck time). Leaving aside the mathematics, Hawking's ingenious way around this can be understood in terms of geometry – specifically, the geometry of a sphere, like the surface of the Earth.

Imagine all of the three dimensions of space represented by a line around the sphere, like a line of latitude. Time can then be represented by a line at right angles to space, measured by the distance from one of the poles of the sphere – say, the North Pole – along a line of longitude. The North Pole itself represents the birth of time, and a tiny circle around the North Pole represents the compact state of the Universe in the Big Bang. As 'time passes', we move to lower latitudes, away from the pole and towards the equator,

2

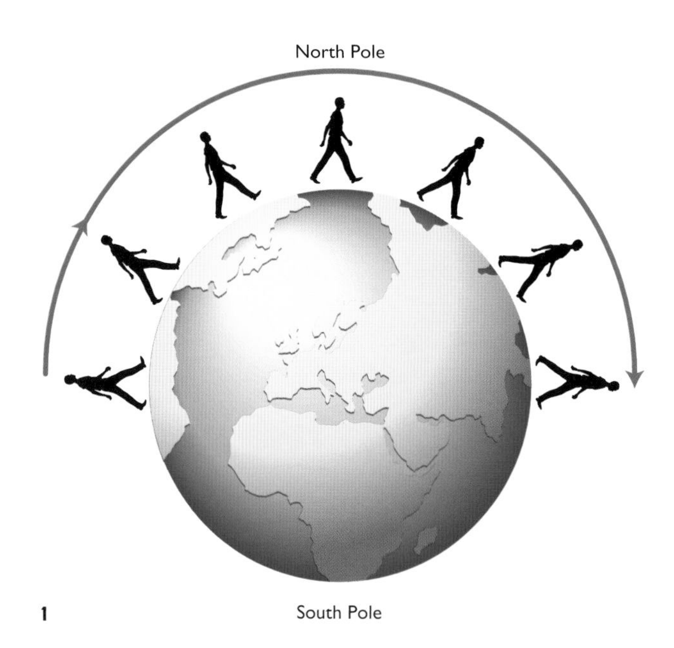

1. If you head due north in a 'straight line' you end up heading south.

and the lines representing space at different times get bigger – the universe expands.

The crucial point is that in this picture there is no edge at the North Pole, just as there is no edge to the Earth at the real North Pole. If you stand at the North Pole, all directions along the surface of the Earth are south. If you stand at the Planck time, all directions of time are into the future. It is as meaningless to ask about what happened 'before' the Big Bang as it is to ask about what exists 'north' of the North Pole.

Hermann Minkowski, who put geometry into Einstein's special theory of relativity in 1908, had been one of Einstein's teachers at university, where he described his pupil as a 'lazy dog', who 'never bothered about mathematics at all.'

Bringing Hawking up to Date

Hawking's original model of the Universe carried this analogy beyond the equator, where the line of latitude is biggest, and down into the 'southern hemisphere', with the lines getting smaller towards the South Pole. The implication was that the universe he described must be closed, and space itself must reach a maximum size and then shrink towards a Big Crunch (▷ p. 99). Hawking even considered the possibility that time might run backwards in the shrinking half of the universe.

But the discovery at the beginning of the twenty-first century that the expansion of the Universe is accelerating because of the lambda term (▷ p. 94) means that although Hawking's model is still a perfectly good one (and might describe a real universe somewhere out there in the Multiverse), it does not describe the Universe we live in. It is easy, though, to adapt Hawking's universe to provide a model like our own Universe. Instead of thinking of spacetime as the surface of a static sphere, it is more appropriate to think of the surface of an expanding balloon. The effect of the lambda term is to make the balloon bigger as time passes, running away so that it becomes impossible for anything starting out from the North Pole ever to reach, let alone cross, the equator.

Everything else in Hawking's model still stands up, and in particular there is still no edge of time at the beginning. The difference is that there is no end of time for a different reason, because of the eternal expansion of the universe. But time is a tricky concept, and some theorists are now exploring the possibility that we should not think of it as flowing from the past into the future at all.

DOES TIME EXIST?

The nature of time is one of the great mysteries in both science and philosophy. We all experience the flow of time, from the past, through the present, and into the future. But where does the past go after the present has happened? And where is the future before the present reaches it? Julian Barbour, an independent physicist based in Oxfordshire, is the latest person to suggest that both past and future really exist 'all the time' (whatever that means) and that all that travels is our conscious awareness of the moment 'now'.

Geography and Relativity

This idea is a natural consequence of the theory of relativity, which sees time as the fourth dimension. Einstein's theory can be expressed in geometrical terms, in which time behaves like a dimension at right angles to the three familiar dimensions of space (up-down, left-right, and forward-back, plus past-future). This is more than a mere analogy. The equations that describe the locations of things in three dimensions of space are an extension of the famous equation derived by Pythagoras regarding the lengths of the sides of right-angle triangles 'the square on the hypotenuse is equal to the sum of the squares on the other two sides'. The equations that describe the locations of things in spacetime are the four-dimensional extension of Pythagoras' equation. And they work perfectly.

This seems to imply that four-dimensional geography is as real as three-dimensional geography or two-dimensional geography. Even if you are somewhere on the surface of the Earth, the Moon is a real place, and you can get to it by travelling through space. In the Einsteinian universe, next Thursday week is always just as real as the Moon, and you get there by travelling through time. The difference is that you have no choice about how fast you travel through time, or where you are going – it's a bit like being on a sealed train travelling at a constant speed through the countryside. At the end of the 1990s, Barbour went further by arguing that there is not even one 'now' (or one 'Thursday week', or one 1914) but an incomprehensible multitude of alternative present moments corresponding to each possible instant of conventional time. He says that the flow of time does not exist, because all times always exist.

Putting the Past in Perspective

Barbour's jumping-off point is the quantum-mechanical vision of reality as an array of parallel worlds, in which every possible outcome of every possible quantum event happens. If I set up an experiment in which an electron has a choice of going through either of two holes, in this picture, the world (the entire Universe) splits into two copies, identical except for the choice of hole the electron goes through. Not very significant – unless I make sure that if the electron goes through hole A it triggers a nuclear bomb which destroys London, while if it goes through hole B London survives. Then, the two alternative realities are quite easily differentiated.

But even this image contains the idea of time flowing, and the Universe(s) evolving. Barbour's Universe (or Multiverse) is utterly timeless. All possible instants in all possible quantum realities always exist. The difference between past and future, in such a picture, is that some instants contain structures that can be called 'records' because they describe accurately what exists in other instants, which we call the past. Those records may be human generated things, like diaries, or natural phenomena, like fossils in geological strata. They define a direction of time, without there needing to be a flow of time. What there is,

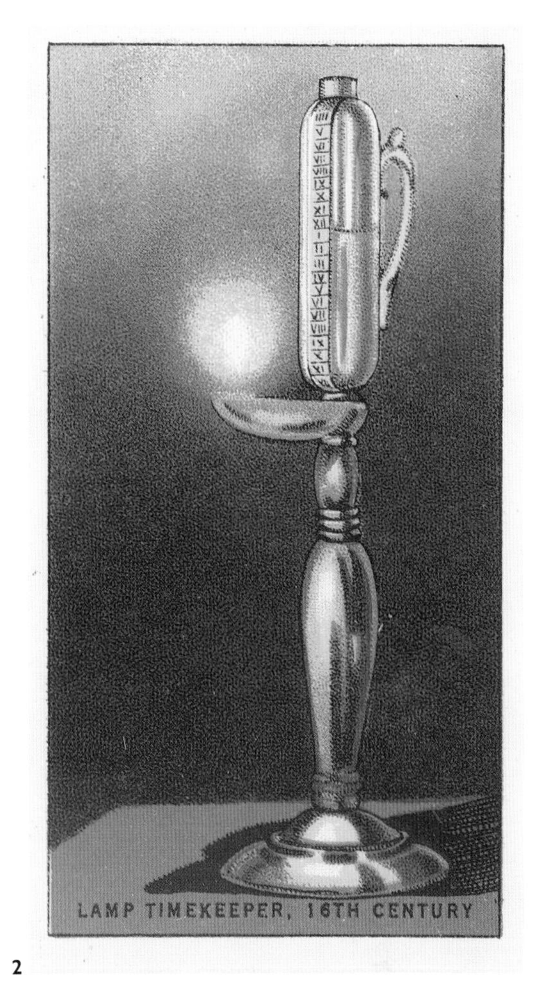

2

2. An early time-keeper recorded the passage of the hours by the slowly falling level of oil in a lamp.

though, is a requirement that the records are self-consistent – the records in each instant describe a possible history.

We perceive a flow of time, says Barbour, because in each instant our brain contains an overlapping set of structures (what he calls a 'time capsule') which gives an illusion of time moving, just as an overlapping set of images from a film strip gives an illusion of motion. But consciousness does not move sequentially along the ranks of time capsules in order – every conscious moment, every time capsule complete with its own historical records and illusion of time passing, exists 'all the time.'

Putting the Worlds Together

There is another way of looking at the many worlds idea, which reduces the number of universes to worry about, and makes some people feel more comfortable about it. David Deutsch, another Oxford physicist (who shares Barbour's view that the flow of time is an illusion), has developed the idea that when a single photon approaches the experiment with two holes, the world does indeed divide in two, with a photon going through one hole in one universe, and through the other hole in the other universe. But when the two possible paths for the photons come back together again in the interference pattern, the two universes merge back together, and that is what makes the pattern. They only exist as separate realities during the time the photon(s) is/are flying through the experiment(s).

According to Deutsch, if we carry out the experiment and allow the interference pattern to form, the splitting and rejoining is a local phenomenon going on in the corner of the lab. But if we look to see which hole the photon goes through, and stop the interference pattern forming, then the world does split permanently into two copies. In a sense, we can make new universes (or new copies of our Universe) on demand. But quantum rules allow universes to be made on a more impressive scale than this.

AN UNCONVENTIONAL SCIENTIST

The outline of Julian Barbour's work scarcely does justice to ideas which are new, mind blowing, and have taken Barbour (right) 30 years to develop. His own book-length exposition of those ideas (*The End of Time*) pulls few punches, and is often an intellectually demanding read. But apart from its intrinsic interest, the story is fascinating because of the way Barbour worked for those 30 years. Although armed with a PhD in physics, he made a conscious decision to opt out of the 'publish or perish' academic rat race, earning his living as a translator of Russian while working on his unconventional ideas with complete freedom. This doesn't mean either that he was isolated or that he is a crank – he regularly attends scientific meetings, publishes papers (in his own good time) and is held in high esteem by his peers.

All this is a shining example of the power of free will and the result of an independent choice of career. Except that Barbour's ideas raise serious doubts about the concept of free will. As he puts it, in the 'many-instants interpretation' of reality, 'each Now' competes 'with all the other Nows in a timeless beauty contest to win the highest probability.' What I *think* this means is that, although everyone reading this book is experiencing a Now in which Julian Barbour is an independent cuss who has come up with a brilliant and novel idea, there are countless more Nows in which countless versions of Julian Barbour took a conventional academic job in 1969 and published a series of dull but worthy papers on conventional topics. Chance, and chance alone, has made 'our' Julian Barbour special.

1. (opposite) Does time really exist? Or is it merely a succession of frozen instants?

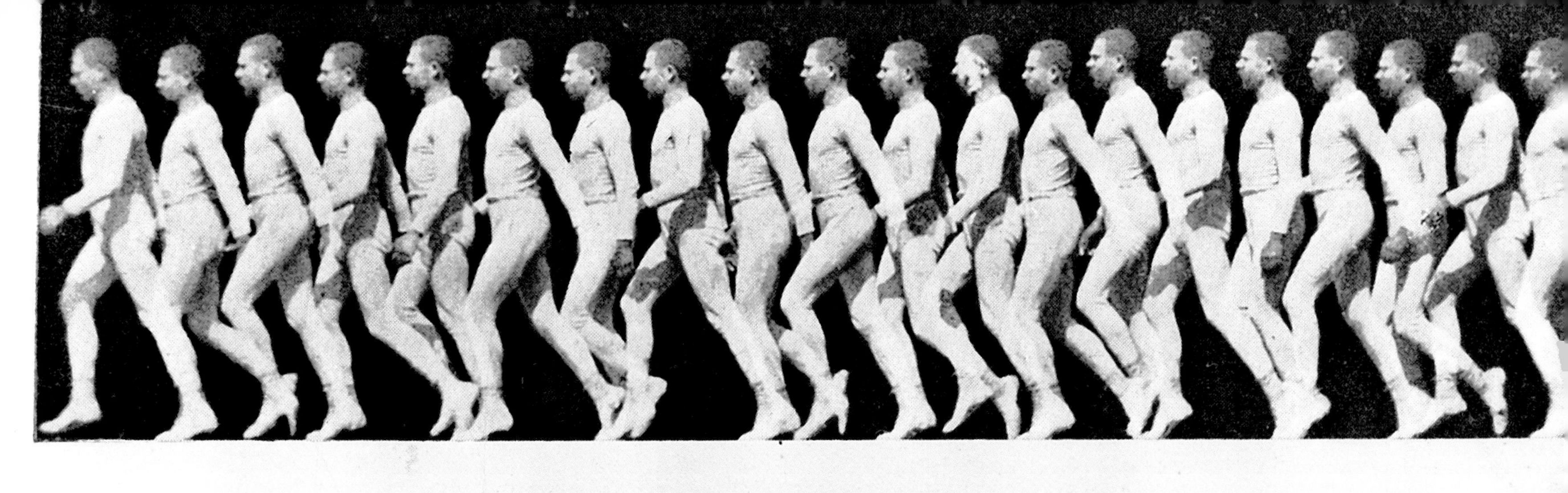

SCHRÖDINGER'S CAT

Erwin Schrödinger, an Austrian physicist, was one of the pioneers who developed quantum mechanics in the 1920s. But by 1935 he was so disgusted with the implications of his own theory that he dreamed up an imaginary experiment to demonstrate its absurdity. It has become the most famous 'thought experiment' in science. A thought experiment is not intended to be carried out 'for real' but it is supposed to have such obvious implications that the result is beyond doubt.

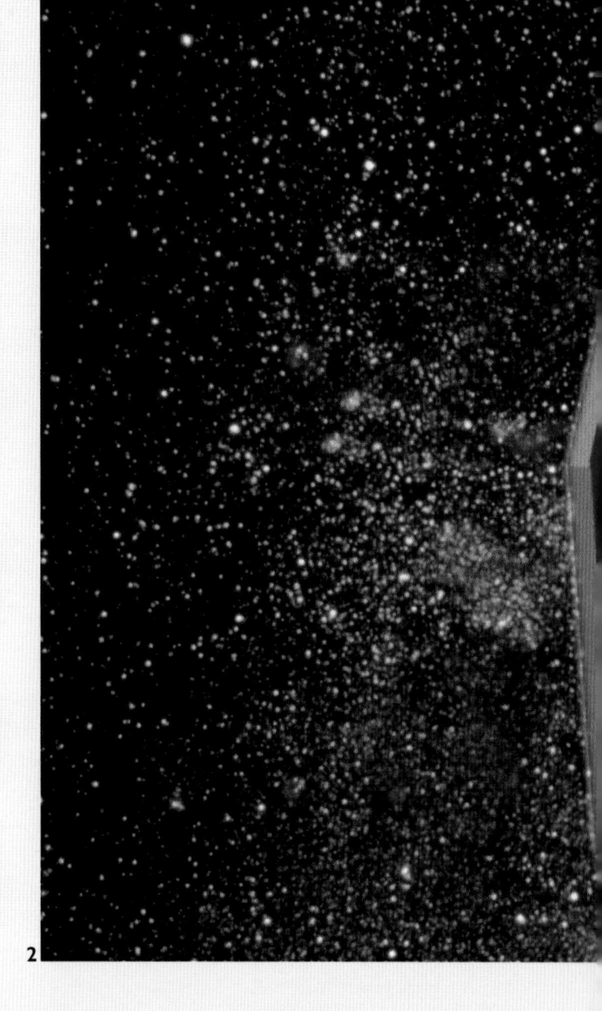

2

1

The Science of Superposition

The basis for Schrödinger's argument is the way that the standard interpretation of quantum mechanics says that a quantum entity exists in a superposition of states until it is measured, and then collapses into a definite state. He used the example of radioactivity, but the 'experiment' works just as well if we think of an isolated electron, which sits in a mixture of two states, spin up and spin down. When the electron spin is measured, it collapses into one of the two states with exactly equal probability. An electron is 'prepared' in such a superposition of states every time one of them is knocked out of an atom, for example, in the electron 'gun' that sends a beam of these particles to paint the picture on a TV screen.

The Cat in the Box

Imagine taking such an electron as it emerges from an electron gun and holding it in a set of magnetic or electric fields, without trying to measure its spin immediately. The electron trap is inside a piece of apparatus connected to a container of poisonous gas, and everything is sealed inside a large room where a healthy cat lives, supplied with plenty of food and water. When the spin of the electron is eventually measured, an automatic device will release the gas and kill the cat if the spin is up, but will let the cat live if the spin is down. Schrödinger pointed out that according to the standard interpretation of quantum mechanics everything in the sealed room, including the cat, is in a 50:50

1. Edwin Schrödinger, inventor of the 'cat paradox'.

2. The Schrödinger's cat thought experiment devised by Edwin Schrödinger (opposite) says that a live cat and its ghost can both exist at the same time.

superposition of states until somebody looks into the room and notices what has happened. The cat is both dead and alive at the same time.

Parallel Possibilities

There are several rival interpretations of quantum mechanics which try to avoid this unwelcome state of affairs. The one that many cosmologists like involves parallel worlds. In this picture, the moment the electron is released, the entire world splits into two copies of itself. In one, the electron has spin down and the cat lives. In the other, the electron has spin up and the cat dies. For a human observer in either world, there is still a 50:50 chance of finding a live cat when you look into the room – but neither cat is ever in a superposition of states.

Extending this example, the entire Universe is multiplied into an infinite number of branches, and anything that can possibly happen does happen in one (or more) of the branches of reality. (This experiment really is 'all in the mind'. Nothing like it has ever been tried with a real cat!)

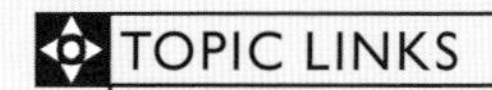

TOPIC LINKS

2.4 The Accelerating Universe
p. 132 Quantum Fluctuations

4.2 A Choice of Universes
p. 205 The Many Worlds
p. 205 The Reality of Duality

4.3 Into the Unknown
p. 230 The Evolution of Universes

4.3 INTO THE UNKNOWN

The idea that alternative worlds might exist somewhere in a quantum Multiverse, combined with the weak anthropic principle, is a neat way of explaining why the Universe we see around us seems to have been designed for life, without invoking a Designer and becoming embroiled in the infinite regress of who designed the Designer, and so on. But all this is rather abstract and philosophical.

The quantum rules do, however, allow for a much more direct relationship between universes, suggesting that they may be physically connected, and that one universe may grow out of another, through a black hole. According to these ideas, Universes reproduce, producing baby universes which grow and have babies in their turn. These ideas come from research at the far frontier of science, and have not yet been tested. But they show where cosmological thinking is taking us in the twenty-first century.

BEYOND THE BLACK HOLE

The equations of the general theory of relativity (a very well established theory that has been tested by many experiments) tell us that the entire contents of a black hole must collapse towards a mathematical point, a singularity (▷ p. 87). The equations of quantum physics (an equally well founded theory) tell us, however, that there is no such thing as a mathematical point, and that nothing can be smaller across than the Planck radius.

Combining these two well-established ideas, physicists conclude that something happens to matter (mass-energy) falling inwards inside a black hole when it reaches the Planck radius. The most likely result is that the material bounces, and expands outwards again. But it doesn't come bursting back out into the universe it fell in from; instead, it is shunted sideways into a new set of dimensions, forming a new expanding universe in its own right.

To put these ideas in perspective, we need to remind ourselves what kind of objects black holes are.

Black Holes Revisited

A black hole is formed by any concentration of matter which has a gravitational field so strong that spacetime is bent right around to form a closed surface. There are two ways this can happen. If any lump of matter is squeezed into a ball so that, although its mass stays the same, its density increases, at some critical density it will become a black hole. This is the kind of black hole left behind by some supernovae. Alternatively, if you keep the density constant and keep adding more mass, at some critical mass it will become a

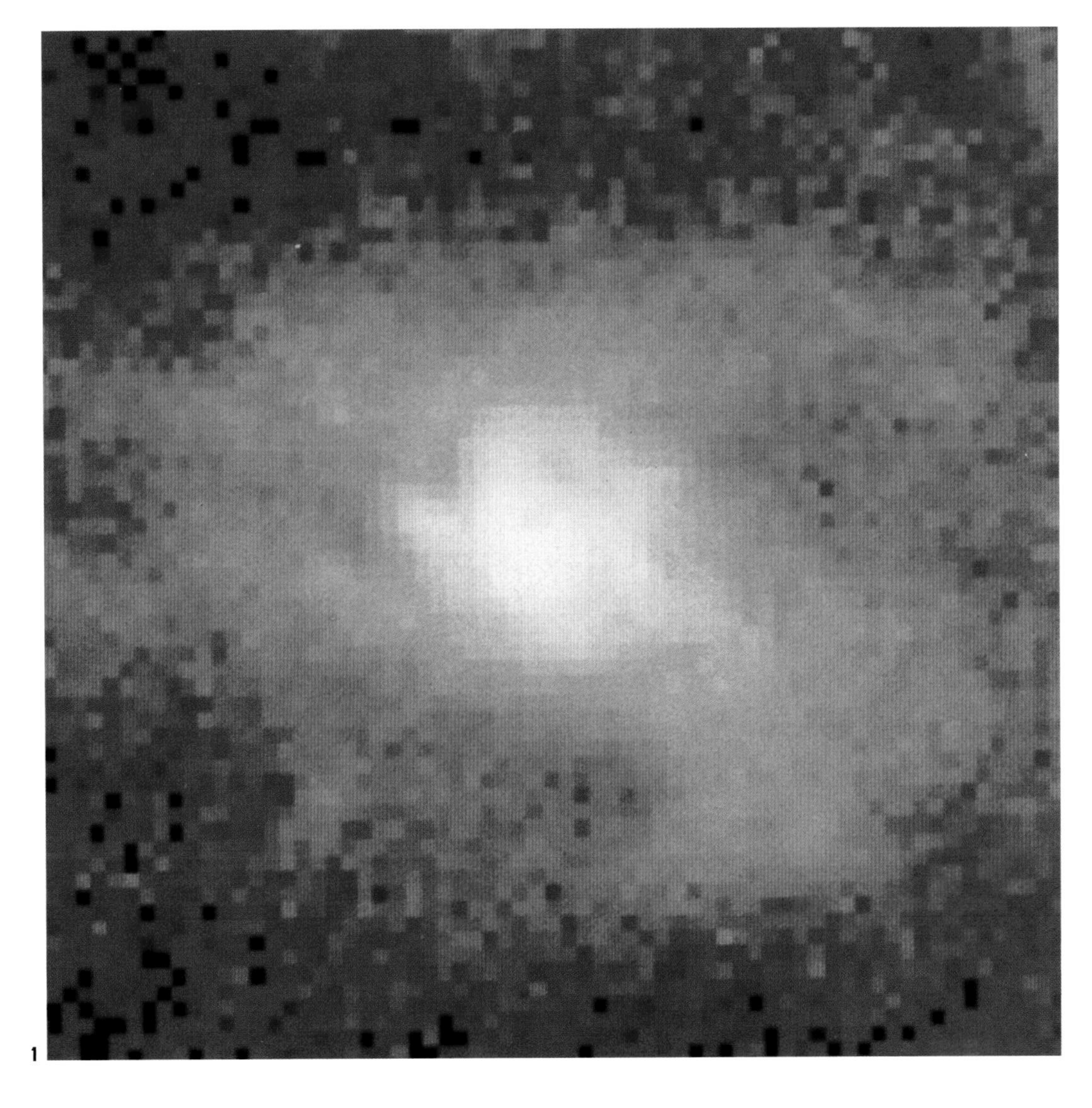

1. A jet of material in the galaxy M87 being shot out by a super-massive black hole at the heart of the galaxy.

black hole. This is the kind of black hole that powers a quasar.

The crucial radius at which an object becomes a black hole is called the Schwarzschild radius, in honour of the German scientist who first realized that the equations of the general theory of relativity predicted the existence of black holes. One way of thinking of the Schwarzschild radius is that the escape velocity from a surface with the Schwarzschild radius is the speed of light. Nothing can escape from inside the Schwarzschild surface, because within it the escape velocity exceeds the speed of light.

To see how the different ways of making black holes work, think of the Sun. If the Sun were squeezed within a ball just 2.9 km across (the Schwarzschild radius for an object with one solar mass), it would become a superdense black hole. But if you could add more matter to the Sun without it collapsing (think of a bag containing a lot of marbles, each representing a star like the Sun), it would become a black hole when it had the mass of a few million Suns, and about the radius of our Solar System – the Schwarzschild radius for an object with a few million solar masses. The density of matter needed to make such a supermassive black hole would only be a little more than the density of water – but the matter would then fall inward towards the centre and be crushed out of existence at the Planck radius.

Einstein's Wormholes

Although black holes were only given their modern name in 1967 (by John Wheeler),

THE MAN WHO INVENTED BLACK HOLES

The first person to suggest the existence of black holes was an eighteenth-century English parson, the Reverend John Michell. But he wasn't just a country parson. Michell, who was born in 1724, became one of the leading scientists of his day before he took Holy Orders, and made his reputation from a study of the disastrous earthquake that struck Lisbon in 1755 (left). He was elected a Fellow of the Royal Society in 1760, and Professor of Geology at the University of Cambridge in 1762. In 1764, he left the university to become the rector of a parish in Yorkshire, but he maintained a keen interest in science and published several important astronomical papers. Among other achievements, he was one of the first people to publish a reasonably accurate estimate of the distance to a star. Using an argument based on its apparent brightness, he calculated that the star Vega is 460,000 times further away from us than the Sun is. In 1783, in a paper read to the Royal Society by his friend Henry Cavendish, Michell pointed out that there could be 'dark stars' in the Universe, so big that light could not escape from them:

'If there should really exist in nature any bodies whose density is not less than that of the Sun, and whose diameters are more than 500 times the diameter of the Sun, since their light could not arrive at us…we could have no information from sight; yet, if any other luminiferous bodies should happen to revolve around them we might still perhaps from the motions of these revolving bodies infer the existence of the central ones.'

These are exactly the kind of black holes now thought to be associated with quasars, and we do indeed infer their existence from the way bright stuff orbits around them.

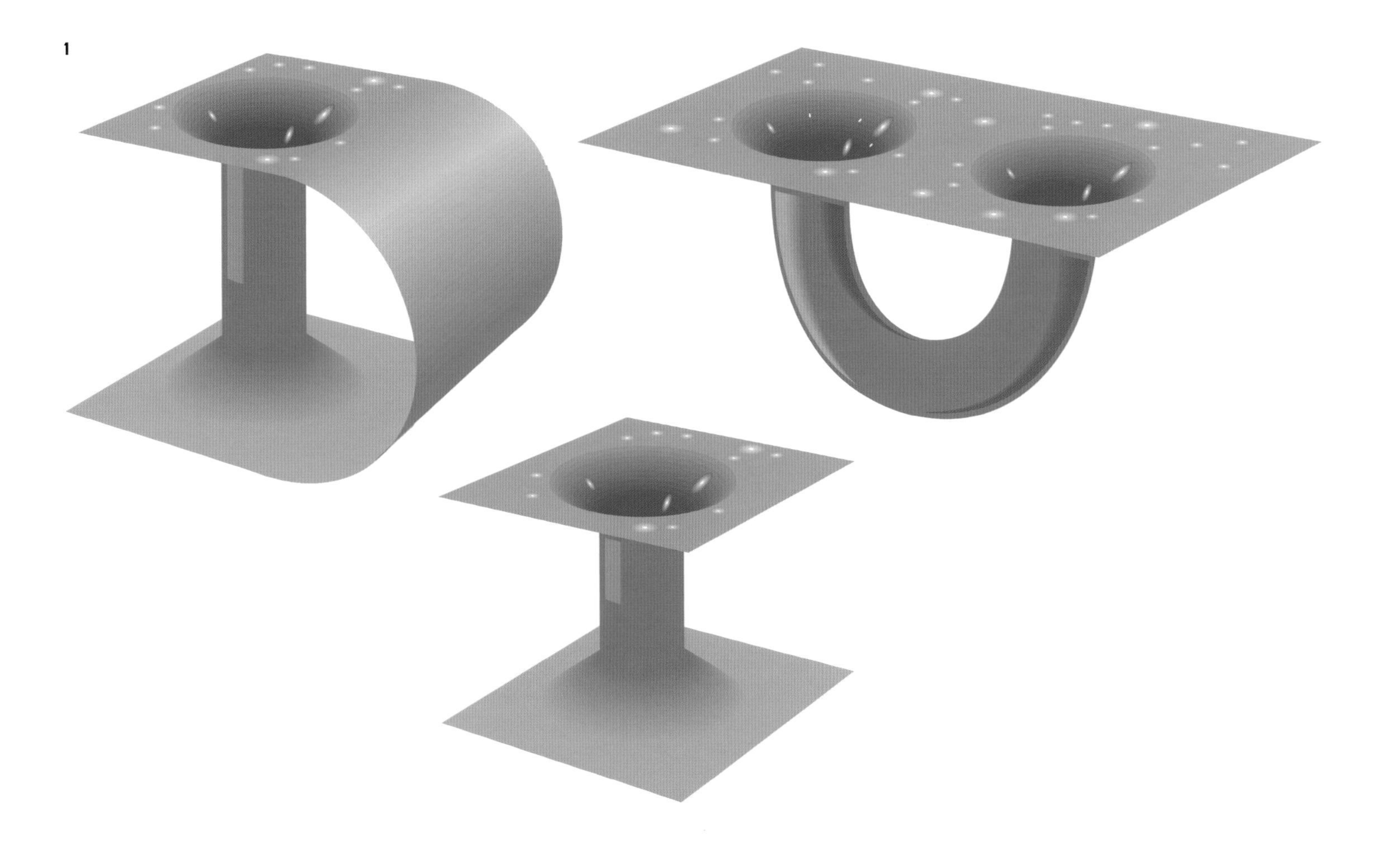

1. 'Wormholes' may join different regions of a single universe, or even separate universes.

Karl Schwarzschild predicted their existence in 1916, and the mathematical description of black holes was studied by Albert Einstein himself in the 1930s. Working with Nathan Rosen, at Princeton University, Einstein discovered that the solutions to the equations of the general theory discovered by Schwarzschild actually described not a single 'hole in space' but a pair of holes connected by a tunnel linking two separate regions of spacetime. This tunnel became known as an Einstein–Rosen bridge, but more recently physicists have taken to calling them 'wormholes'.

An Einstein–Rosen bridge, or wormhole, joins two black holes in different locations in spacetime. Originally, people thought that if such entities were real they could link different parts of our Universe, like a cosmic subway. This is still true; but because space*time* is involved not just space, a wormhole might also, in principle, link two different times in our Universe – it could be a kind of time machine. And now cosmologists speculate that if other universes exist, then wormholes may provide links between different universes.

Quantum Foam

Don't get too excited, however, about the possibility of travelling through a wormhole to some other place, or time, or some other universe. It would be extremely difficult (for all practical purposes, it may be impossible) to build a large wormhole through which people could travel. Natural wormholes, if they occur at all, exist on the scale of the Planck length, whatever the size of the Schwarzschild radius of the black hole that acts as the gateway to a particular cosmic subway.

Physicists who study the mathematics of such entities are still intrigued by wormholes, however, because quantum processes operating on the Planck scale may produce vast numbers of tiny (sub-sub-microscopic) wormholes, and these could provide the structure of spacetime itself. The quantum wormholes would be like the strands of a carpet, woven together to make the seemingly solid structure of spacetime (the carpet) itself. And they could be the seeds of baby universes.

1

1, 2 and 3 (opposite): The surface of the sea looks less and less smooth the closer you get to it.

2

HOW TO MAKE A UNIVERSE

The idea that spacetime may be woven together out of quantum-scale wormholes is related to another of John Wheeler's suggestions, which portrays spacetime as a 'foam' of quantum entities, popping in and out of existence at the scale of the Planck length. This could include black hole pairs and the wormholes that connect them. An analogy Wheeler makes is with the appearance of the surface of the sea. From a high-flying aircraft, the sea looks smooth and flat. Closer up, it looks rough, and closer still the surface dissolves into a constantly changing foam of bubbles and tiny waves. Spacetime may only seem smooth to us because we are so much larger than the Planck scale – remember that the Planck length, 10^{-33} cm, is 10–20 times the size of a proton.

Virtual Reality

Wheeler's idea is based on the prediction from quantum mechanics that quantum entities such as pairs of particles can appear out of nothing at all for a short time ('virtual pairs'). We met this idea in Chapter 2, in the

1. A pair of interacting galaxies known as the antennae.

2. A gamma ray burster pictured by the Hubble Space Telescope.

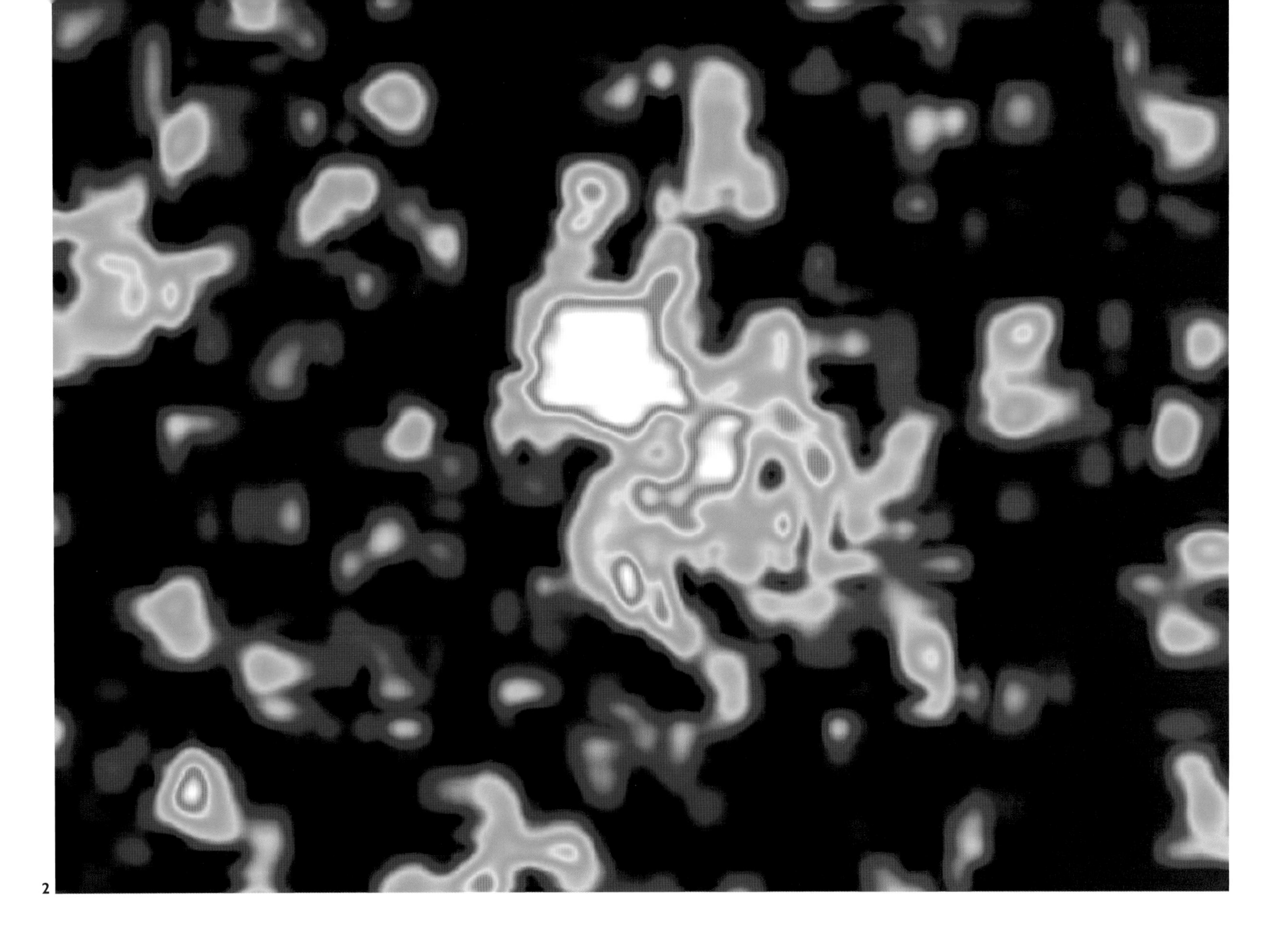

2

context of the term that drives the acceleration of the Universe. But this is not just a speculation floated by the cosmologists to explain their observations of distant galaxies. Such quantum fluctuations are an integral part of quantum mechanics, and give a structure to the vacuum (to 'empty space'), with consequences that can be measured in the laboratory.

Particles made out of nothing in this way always come in pairs (called virtual pairs) with opposite quantum properties. For example, a virtual electron is always accompanied by a virtual positron, a particle like an electron but with positive charge rather than negative charge. Close to a real charged particle, such as a permanent electron, even during the brief lifetime of the virtual pair there will be time for the virtual positron to move towards the real electron and the virtual electron to be repelled from it. This produces a shielding effect around the real electron, and reduces the effect of its charge, as measured by the influence on another real electron. The consequences of this shielding can be predicted by quantum theory, and exactly match the measured behaviour of charged particles such as electrons. There is no doubt that quantum fluctuations are real.

Something For Nothing

There is nothing in the quantum rules, however, which says that only particles can be created temporarily out of nothing at all

Astronomers estimate there are at least 100 million black holes in our Galaxy. With over 100 billion galaxies, that means our Universe may already have had ten thousand million billion (10^{19}) offspring.

THE NEGATIVITY OF GRAVITY

The idea that gravity corresponds to negative energy is one of the hardest scientific concepts for a non-specialist to come to terms with. But it is also one of the most important, for without the negativity of gravity the Universe probably would not exist.

1. It takes a lot of energy to lift anything out of the gravitational well of the Earth.

Extending to Infinity

Without delving into the equations of the general theory of relativity, the best way to get a handle on this is to imagine dis-assembling something (anything!) and dispersing its component parts to infinity. This is only a 'thought experiment', so you can do this at the level of atoms, or particles such as protons and neutrons, but the analogy works just as well for a pile of bricks.

Remember that gravity obeys an inverse square law. So the force of gravity between any two bricks is proportional to one divided by the square of the distance between them. If the bricks are infinitely far apart, the force must be zero, because anything divided by infinity (let alone by infinity squared) is zero.

Storing Energy

The amount of energy that is stored in a gravitational field depends on the force between the objects involved. This isn't only true for gravity. Think of a spring. When the spring is relaxed, loose and floppy, it stores no energy (except, of course, its mc^2). When you stretch it, the force pulling the spring inwards gets bigger, and more energy is stored in the spring. But with gravity, there is *less* force when things are stretched apart. If you could stretch a spring to infinity without breaking it, there would be a huge force pulling it inward, and it would store a huge amount of energy. With gravity, though, there is a huge force when things are very close together. When

the components are at infinity, there is zero force, so there is zero energy in the gravitational field.

Extracting Energy

But we know that when things fall together, gravitational energy is released. That's how stars get hot enough inside to start burning their nuclear fuel. Imagine all your bricks, dispersed to infinity, given a gentle nudge so that they start falling together. As they fall, they release energy. But they start out with zero energy – so after releasing some, they are left with less than nothing. It's like writing a cheque when you have no money in your bank account – suddenly you have *negative* amounts. In all real objects, the energy of their gravitational field is negative.

If your bricks fall all the way from infinity down to a point (or the Planck length), then, the equations of the general theory tell us, the amount of energy released is equal and opposite to their total mass-energy, $-mc^2$. Turning this around, a universe destined to expand to infinity can appear out of nothing at all at the Planck scale, because its mass-energy is precisely cancelled out by its negative gravitational energy.

2. The Lagoon nebula, about 3500 light years away from us, is a stellar nursery where thousands of new stars have recently been formed.

TOPIC LINKS

1.2 Star Birth
p. 31 Turning up the Stellar Heat

1.4 Out with a Bang
pp. 72-73 The Search for Black Holes

2.1 The Big Bang
p. 87 The Singular Beginning

2.4 The Accelerating Universe
p. 132 Quantum Fluctuations

4.3 Into the Unknown
p. 227 Einstein's Surprise

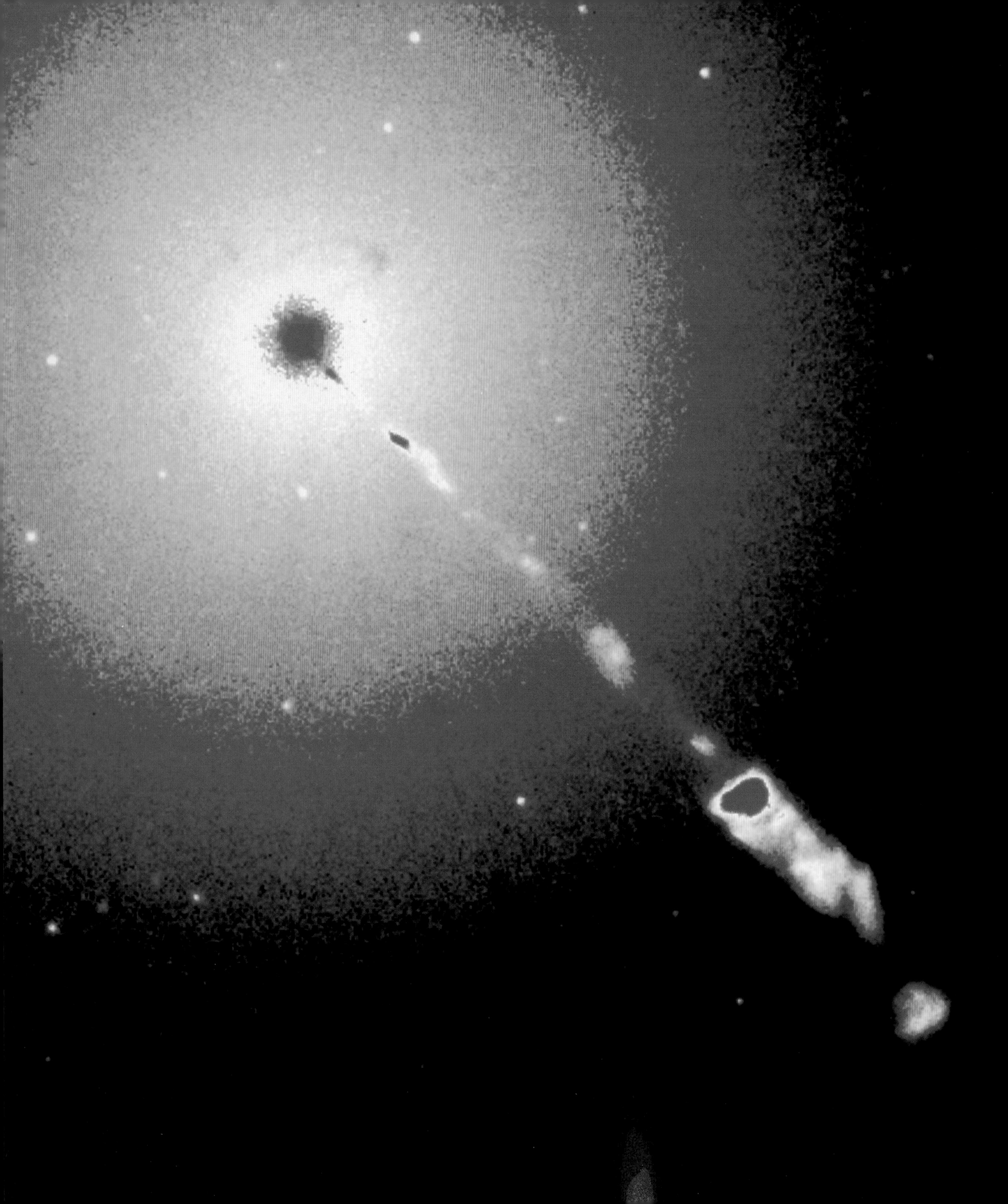

1. (opposite) Close-up view of the jet of gas being emitted from the black hole at the centre M87.

in this way. Little packets of pure energy (bubbles with a radius roughly equal to the Planck length) can also appear out of nothing at all, provided that they disappear within the time allowed by Heisenberg's uncertainty relation. But that time can be very long indeed.

Remember that the less energy a quantum fluctuation contains, the longer it can live. The energy of a little packet of mass-energy actually comes in two forms – from its mass, but also from its gravity. Curiously, the energy of a gravitational field is actually negative. This means that for a bubble with just the right amount of mass-energy the two effects cancel out, and the overall energy of the bubble is zero. In that case, it could last forever. 'Just right' means that the bubble contains just enough mass to make it flat, on the edge of being a black hole. If our Universe is flat, it also contains zero energy overall.

As long ago as 1973, Edward Tryon, of City University in New York, pointed out that as far as quantum physics was concerned the entire Universe could be nothing more or less than a quantum fluctuation of the vacuum.

Blowing Bubbles

The big snag with Tryon's idea was that quantum fluctuations occur on the Planck scale. There was, indeed, nothing in the

EINSTEIN'S SURPRISE

The idea that the negativity of gravity can precisely cancel out the mass-energy $E=mc^2$ in a lump of matter (or the Universe) is so surprising that even after you have had it explained to you, or studied the equations for yourself, it is hard to believe. If you feel that way, you are in very good company.

The first person to realize this possibility was the physicist Pascual Jordan, working in the United States in the 1940s. At that time, Albert Einstein (right) worked as a consultant for the US Navy, assessing schemes for new weapons sent to the military by well-meaning civilians. Einstein was very good at this – in the first decade of the twentieth century he had worked as a Patent Officer in Switzerland, so he was used to finding the flaws in inventions. Every fortnight or so, George Gamow, who was based in Washington DC and also involved in war work, would take the latest batch of ideas up to Einstein in Princeton.

On one of these visits, as Gamow later recalled in his autobiography *My World Line*, the two physicists were walking from Einstein's home to the Institute for Advanced Study, where Einstein worked, when Gamow casually mentioned that Jordan had told him that a star could be made out of nothing at all, because at the point of zero volume its negative gravitational energy would precisely cancel out its positive mass-energy. 'Einstein stopped in his tracks,' Gamow tells us, 'and, since we were crossing a street, several cars had to stop to avoid running us down.'

quantum rules to forbid a bubble containing as much mass-energy as our entire Universe to appear on the Planck scale. But it would have an enormous gravitational pull, which would crush it out of existence immediately. Tryon's idea lay dormant until the 1980s, when inflation theory provided the mechanism which can take a superdense quantum seed and blow it up to the size of an observable object in our world. The enormous push given by inflation acts like antigravity, flattens spacetime, and prevents the baby universe crushing itself out of existence as soon as it forms.

There is no reason, however, to stop at one universe. In this picture, universes of all sizes should be produced by quantum fluctuations, some with just enough strength to expand a little way, others where the term is powerful enough to keep the universe expanding forever, and with all possible intermediate values. Each universe is like another bubble in the foam of spacetime, expanding in its own way with no contact with its neighbours, and we are back to the weak anthropic principle to explain why the Universe we live in is as it is.

But Tryon's ideas, inflation, and the mathematics of black holes can be combined to give a more intriguing insight into how we got here.

Black Holes and Baby Universes

The seed of a universe like our own could come into existence as a concentration of mass-energy containing all the mass-energy of the observable Universe in a volume with the Planck radius, 'born' at an age of 10^{-43} seconds, the Planck time (▷p. 98). Inflation could drive the expansion of such a baby universe up to a large size, and this could be happening anywhere in the vacuum of space around us (or, indeed, in the space between the atoms of the page you are now reading). If so, these other universes wouldn't explode outwards in your face, filling up our own region of spacetime, but would exist in their own sets of dimensions, all of them at right angles to the dimensions of our spacetime. But it would also be possible, in principle, to make a universe, deliberately.

Universes on Demand

Several physicists, including Alan Guth, one of the pioneers of inflation theory, have explored this possibility mathematically. One of the key points in their argument is that even in order to make as large a Universe as the one we live in, you do not need a lot of mass-energy to start with. Because of the negativity of gravity, nature can make universes out of nothing at all. We cannot quite do that, because we would have to put energy in to trigger the process of inflation. But the amount of energy required is surprisingly small, compared even with the output of a star like the Sun.

The requirement to kick-start inflation is a temperature equivalent to about 10^{24} K and a very high density. The energy required to produce these conditions is no more than could be produced by a few hydrogen bombs; the problem is confining that energy, even if only for a split-second, in a tiny volume of space, about the size of an atom. If this could be done, the equations studied by Guth and others show that under some circumstances inflation will occur in the compressed region.

1. The technology of the hydrogen bomb almost gives us the ability to make black holes – and perhaps baby universes.

There is another way to make a universe, equally impractical as far as present-day technology is concerned, but allowed by the laws of physics, and that is to build a black hole. The stuff inside a black hole must, as Roger Penrose proved nearly four decades ago, fall towards a singularity. At the place where the infalling matter is squeezed into a Planck volume, quantum processes take over and shunt the infalling stuff sideways into a new set of dimensions, forming a new expanding universe. Once again, however, it doesn't matter how much or how little mass-energy goes into the black hole, or whether it is in the form of water, or hydrogen, or peanuts, or anything else. Thanks to the negativity of gravity, a universe as impressive as our own can be made from a black hole with any amount of mass.

But in another twist to the tale, there is nothing to say that the laws of physics need be exactly the same in the baby universe as in the universe which gave it birth.

1

THE EVOLUTION OF UNIVERSES

Lee Smolin, of Pennsylvania State University, has come up with one of the most spectacular speculations in science. He has made the guess (and as yet it is no more than a guess) that every time a baby universe buds off from another universe through a black hole/wormhole 'umbilical cord', the laws of physics in the new universe are slightly different (but not dramatically different) from the laws in the parent universe. This would echo the way in which children are slightly different from their parents, but horses always give birth to horses, cats to cats, and so on.

Provided this is the case, when new universes are produced by quantum fluctuations, they will come in all varieties, some expanding more than others. But when new universes are born out of old universes, they will have a family resemblance to their parent. If a quantum fluctuation doesn't grow big enough to have very many black holes, it will leave few offspring (perhaps none at all). But the bigger a universe grows, the more black holes it will produce, and the more babies it will have – crucially, those babies will share the propensity of the parent universe to grow big and have lots of babies in their turn.

Smolin sees this as leading inevitably to the production of big universes in which the laws of physics are fine-tuned to make the maximum number of black holes. Universes like our own should, in this picture, actually be the most common kind of universes in the Multiverse, and we don't even need to invoke the weak anthropic principle to explain why we live in such a universe. But if the Universe is the result of a kind of evolutionary process that favours the production of black holes, why are we here at all?

1. The young star Herbig-Haro 32, surrounded by the nebula in which the star formed.

2. The giant molecular cloud M42.

2

People are Parasites

Smolin's answer is that we are at best a by-product of the processes that produce black holes, and at worst parasites on the Universe. This happens because size isn't the only thing that matters if a universe is to have lots of babies – the laws of physics have to be just right for making black holes in that universe.

The most efficient way to turn matter into new universes is to have lots of small black holes, because you get one new universe for each black hole. A hundred million black holes each with the mass of the Sun are much better (a hundred million times better) than a single black hole with a hundred million solar masses. Stellar mass black holes are, of course, made from stars, as part of the natural cycle of starbirth and stardeath. And that natural cycle depends crucially on the presence of elements such as carbon and oxygen, as one example illustrates.

When a cloud of gas and dust (a giant molecular cloud) starts to collapse to form stars, it gets hot inside. As the first stars form, they radiate a lot of ultraviolet radiation and visible light, which tends to blow the cloud apart, and this would stop stars like the Sun forming unless the cloud could get rid of the heat. It does so because compounds such as carbon monoxide (CO) and water (H_2O) absorb the ultraviolet and visible light, and re-radiate it at infrared wave-lengths which can penetrate the dusty cloud and escape into space. This allows the cloud to collapse and produce many more stars, includ-ing supernovas, the birthplaces of black holes.

Without the carbon and oxygen, only a few stars would form (but at least a few would form, which is how the first carbon and oxygen was made in our Universe). As we have seen, the laws of physics seem to be fine-tuned to allow carbon and oxygen to be formed inside stars, and this benefits carbon-based, oxygen-breathing life forms like ourselves. But if Smolin is right, the 'coincidences' that allow carbon and oxygen to exist have happened neither by accident nor by design, but as part of a process of natural selection among universes. Carbon and oxygen are an integral part of the reproductive cycle of the Universe, and we just go along for the ride.

It is a sobering removal of humankind from the centre of the cosmic stage – but the upside is that if these ideas are correct, there are countless other universes, all relatively rich in carbon and oxygen, where organic beings like ourselves can thrive.

Further information

Aczel, Amir, *Probability 1* (Harcourt Brace, 1998). Arguing the case that we are not alone in the Universe.

Christianson, Gale, *Edwin Hubble* (Farrar, Straus & Giroux, 1995). Definitive biography of the man who discovered that the Universe is expanding.

Crick, Francis, *Life Itself* (Macdonald, 1982). Nobel prizewinner discusses the puzzle of the origin and nature of life.

Croswell, Ken, *Planet Quest* (Free Press, 1997). The most accessible account to date of the discovery of alien 'solar systems'.

Gamow, George, *Mr Tompkins in Paperback* (Cambridge UP, 1967). Classic stories of the dream world in which Mr Tompkins experiences the mysteries of the Universe. An updated version by Russell Stannard is also available.

Gribbin, John, *In Search of Schrödinger's Cat* (Black Swan, 1984). Classic explanation of the 'many worlds' interpretation of quantum mechanics.

Gribbin, John, *The Birth of Time* (Phoenix, 1999). Up-to-date discussion of the evidence for the Big Bang.

Gribbin, John, *Stardust* (Allen Lane, 2000). The most accessible account of the way the elements were manufactured inside stars.

Henbest, Nigel, and Couper, Heather, *The Guide to the Galaxy* (Cambridge UP, 1994). Lavishly illustrated and accurate introduction to the geography of the Milky Way.

Longair, Malcolm, *Our Evolving Universe* (Cambridge UP, 1996). An overview of cosmology from one of Britain's leading astronomers.

Malin, David, *The Invisible Universe* (Bulfinch Press, 1999). Large format collection of spectacular astrophotographs of objects that cannot be seen by the naked eye.

Mallove, Eugene, and Matloff, Gregory, *The Starflight Handbook* (Wiley, 1989). Describes in detail the prospects for interstellar flight.

Murdin, Paul and Lesley, *Supernovae* (Cambridge UP, 1985). A clear account of what goes on when stars explode.

Petersen, Carolyn, and Brandt, John, *Hubble Vision* (Cambridge UP, second edition,1998). Spectacular images from the Hubble Space Telescope, with accompanying explanatory text.

Rees, Martin, *Just Six Numbers* (Weidenfeld & Nicolson, 1999). Britain's Astronomer Royal looks at the cosmic coincidences which define the Universe and life.

Ridpath, Ian, *Book of the Universe* (Dragon's World, 1991). Excellent beginner's guide to the cosmos.

Smolin, Lee, *The Life of the Cosmos* (Weidenfeld & Nicolson, 1997). Controversial but intriguing account of the 'baby universes' hypothesis.

Shklovskii, I. S., and Sagan, Carl, *Intelligent Life in the Universe* (Holden-Day, 1966). Still the best book about the search for ET.

Weinberg, Steven, *The First Three Minutes* (Deutsch, 1977). Nobel prizewinner discusses the puzzle of the origin and nature of the Universe.

Zuckerman, Ben, and Hart, Michael (editors), *Extraterrestrials: Where Are They?* (Cambridge UP, second edition, 1995). Is there intelligent life in the Universe anywhere except on Earth?

WEBSITES:

NASA websites:
www.nix.nasa.gov, www.photojournal.jpl.nasa.gov

Views of the Solar System
www.solarviews.com

Society of Popular Astronomy
www.popastro.com

Sky and Telescope Magazine
www.skypub.com

Hubble Space Telescope
www.stsci.edu

Satellite predictions
www.heavens-above.com

Picture credits

BBC Worldwide would like to thank the following for providing photographs and for permission to reproduce copyright material. While every effort has been made to trace and acknowledge all copyright holders, we would like to apologise should there have been any errors or omissions.

AKG London pages 127, 131, 227; **Brian and Cherry Alexander** page 209; **Allsport** page 93 *above*; **Anglo-Australian Observatory** pages 15, 20 *below*, 21, 23, 32 *right*, 40, 66, 68, 81 *right*, 120, 222; **Art Archive** pages 65, 193; **AT & T Bell Laboratories** page 88; **BBC Worldwide Ltd** pages 36, 42, 51 *above*, 54, 58, 69, 70 *above*, 70 *below*, 71, 82 *below*, 95, 102 *above*, 115, 117, 145 *below*, 148, 156, 176, 194 *left*, 200, 208, 210, 219; **Boomerang Collaboration** pages 134 *left*, 134 *right*, 135; **British Film Institute** page 155; **Corbis** page 124; **European Space Agency** page 16; **Galaxy Picture Library** pages 17, 20 *above*, 22, 26, 27, 28 *below*, 29 *above*, 34, 35, 45, 49, 56 *left*, 56 *right*, 61, 74, 79, 91, 111, 113, 121, 129, 147 *below*, 151, 154, 169, 171, 175, 217, 223, 231; **Genesis Space Photo Library** pages 76, 80, 143, 144, 145 *above*, 224, 232; **Hulton Getty Picture Collection** pages 153, 218; **Imagebank** pages 35 *below*, 199; **Images Colour Library** pages 29 *below*, 33, 102 *below left*, 185 *above left*, 204; **Mary Evans Picture Library** pages 75, 98 *above*, 187, 211; **Mount Wilson Observatory** pages 128; **NASA** pages 14 *right*, 32 *left*, 52, 53, 67, 132, 142 *below*; **Oxford Scientific Films** pages 81 *left*, 119, 146, 173, 180 *top*, 180 *below left*, 180 *below right*, 185 *above right*, 185 *below*, 220 *below*; **Pictor International** pages 112 *left*; **Popperfoto** page 229; **Princeton University** page 206; **Robert Harding Picture Library** pages 103 *left*, 207, **Rockefeller University Archives** page 192, **Royal Society** page 50, **Science Photo Library** pages 2, 5, 8, 11, 11 *right*, 13, 18 *left*, 18 *right*, 19 *below*, 24, 25, 28 *above*, 31, 39 *above*, 39 *below*, 43, 44, 44 *right*, 46, 51 *below*, 55, 57, 59, 63, 64, 72 *above*, 72 *below*, 73 *left*, 73 *right*, 82 *above*, 83, 84, 86, 89 *left*, 89 *right*, 90, 93 *below*, 94, 96, 97, 98 *below*, 99, 100, 100 *below*, 101 *above left*, 101 *above right*, 101 *below left*, 101 *below right*, 104, 107 *left*, 107 *right*, 108, 109, 110 *above*, 110 *below*, 112 *right*, 114, 116, 118, 122, 123, 125, 130, 133, 136, 139, 141, 142 *above*, 147 *above*, 149, 155, 157, 158, 159 *left*, 159 *right*, 160, 161, 162, 163, 164 *left*, 164 *right*, 165, 166, 170, 172, 177, 181, 182, 189, 190, 194 *right*, 195, 203, 205, 212, 214, 215, 220 *above*, 221, 225, 226, 230, 234; **Science and Society Picture Library** pages 141, 213, **Search for Extraterrestrial Intelligence Institute** page 174; **Shout Picture Library** page 103 *right*; **Telegraph Colour Library** pages 102 *below right*, 186, 197, **University of Colorado** page 85.

Index